CONTENTIOUS GEOGRAPHIES

Contentious Geographies
Environmental Knowledge, Meaning, Scale
Edited by
MICHAEL K. GOODMAN, King's College London, UK
MAXWELL T. BOYKOFF, University of Oxford, UK
KYLE T. EVERED, Michigan State University, USA

The human-environment relationship – intimately intertwined and often contentious – is one of the most pressing concerns of the 21st century. Explored through an array of critical approaches, this book brings together case studies from both the global North and South to present significant cutting-edge research into political ecologies as they relate to multi-form contestations over environments, resources and livelihoods.

Contentious Geographies examines the conflict-ridden governance of the environment – and of its contested social and material constructions – through specific references to science, policy, media and technology. Contributing chapters interrogate the multiple ways that knowledge and epistemic framings embody power through discursive and technological regimes and translate these into political economic and ecological outcomes. Covering a range of issues, such as popular discourses of environmental 'collapse', climate change, water resource struggles, displacement, agro-food landscapes, and mapping technologies, this edited volume works to provide a broad and critical understanding of the narratives and policies more subtly shaping and being shaped by underlying environmental conflicts.

By exploring the power-laden processes by which environmental knowledge is generated, framed, communicated and interpreted, Contentious Geographies works to reveal how environmental conflicts can be (re)considered and thus (re)opened to enhance efforts to negotiate more sustainable environments and livelihoods.

ASHGATE STUDIES IN ENVIRONMETAL POLICY AND PRACTICE
Series Editor: Adrian McDonald

Based on the Avebury Studies in Green Research series, this wide-ranging series still covers all aspects of research into environmental change and development. It will now focus primarily on environmental policy, management and implications (such as effects on agriculture, lifestyle, health etc), and includes both innovative theoretical research and international practical case studies.

Contentious Geographies
Environmental Knowledge, Meaning, Scale

Edited By

MICHAEL K. GOODMAN
King's College London, UK

MAXWELL T. BOYKOFF
University of Oxford, UK

KYLE T. EVERED
Michigan State University, USA

Routledge
Taylor & Francis Group

LONDON AND NEW YORK

First published 2008 by Ashgate Publishing

2 Park Square, Milton Park, Abingdon, Oxon OX14 4RN
711 Third Avenue, New York, NY 10017, USA

Routledge is an imprint of the Taylor & Francis Group, an informa business

First issued in paperback 2016

British Library Cataloguing in Publication Data
Contentious geographies : environmental knowledge, meaning,
  scale. - (Ashgate studies in environmental policy and
  practice)
  1. Environmental responsibility 2. Natural resources -
  Social aspects 3. Resource-based communities 4. Nature -
  Effect of human beings on 5. Environmental policy - Social
  aspects
  I. Goodman, Michael K., 1969- II. Boykoff, Max III. Evered,
  Kyle
  363.7

Library of Congress Cataloging-in-Publication Data
Goodman, Michael K., 1969-
 Contentious geographies : environmental knowledge, meaning, scale / by Michael K. Goodman, Max Boykoff and Kyle Evered.
   p. cm. -- (Ashgate studies in environmental policy and practice)
 Includes bibliographical references and index.
   1. Environmental responsibility. 2. Natural resources--Social aspects. 3. Resource-based
communities. 4. Nature--Effect of human beings on. 5. Environmental policy--Social aspects.
I. Boykoff, Max. II. Evered, Kyle. III. Title.

 GE195.7.G66 2008
 333.72--dc22

                                                                        2007046575

 ISBN 13: 978-0-7546-4971-7 (hbk)
 ISBN 13: 978-1-138-27559-1 (pbk)

Contents

List of Figures

List of Maps

List of Tables

Notes on Contributors

Trevor L. Birkenholtz is an Assistant Professor in the Department of Geography, Rutgers University. He is a cultural and political ecologist with interests in development policy, water resources management and nature-society-technology relations, particularly in South Asia.

Maxwell T. Boykoff is a James Martin 21st Century Research Fellow in the Environmental Change Institute at the University of Oxford. He holds a PhD in Environmental Studies (with a parenthetical notation in Sociology) from the University of California, Santa Cruz, and Bachelor of Sciences from the Ohio State University. Max has analyzed how various non-state actors influence environmental science, policy and practice. His oft-collaborative research has examined media coverage of climate change, how certain discourses influence environmental policy considerations, the role of celebrity endeavours in climate change issues, and links to ethics, environmental justice movements and public understanding.

Jessica Budds is a Lecturer in Geography within the Faculty of Social Sciences at the Open University. She works on environment and development issues, focusing on the political ecology of water resources management, with particular reference to Latin America. Her recent research has explored the socio-ecological outcomes of the application of market approaches to water sector policies, through urban water privatisation in the global South, tradable private water rights in Chile, and conflicts over water in relation to the liberalization of mining in Peru.

Kyle T. Evered received his PhD in Geography (2002) from the University of Oregon and was an Assistant Professor of Geography at Illinois State University for three years (2002–2005); he is currently an Assistant Professor of Geography and is on the Advisory Committee for the Muslim Studies Program at Michigan State University. His broad research interests include the political ecologies of agriculture and wetland ecosystems in Turkey, the Middle East and Eurasia, and the topics of identity, place, and regionalization in Turkey and Eurasia. Current research projects in Turkey involve the political ecologies of opium poppy production, rose cultivation and marketing, and the impacts of the European Union and neo-liberalization on Turkish farmers.

Tim Forsyth is a Reader at the Development Studies Institute, London School of Economics. He has written on various themes of science-policy within public policy and development, including *Critical Political Ecology: The Politics of Environmental Science* (2003, Routledge), and *Forest Guardians, Forest Destroyers: The Politics of*

Environmental Knowledge in Northern Thailand (2008, University of Washington Press, with Andrew Walker). These works aim to show how truth claims within environmental politics reflect political influences, and aim to develop new forms of environmental governance that can make environmental policy more effective and inclusive.

Jefferson Fox is the Coordinator of Environmental Studies and a Senior Fellow at the East-West Center in Honolulu. He received his PhD in Development Studies from the University of Wisconsin-Madison in 1983. His research focuses on land-use and land-cover change in Asia and the possible cumulative impact of these changes on the region and the global environment. He has co-edited several books, most recently, *People and the Environment: Approaches for Linking Household and Community Surveys to Remote Sensing and GIS* (2003, Kluwer Academic Press). His ongoing research includes understanding dynamic resource management systems and land cover transitions in Montane Mainland Southeast Asia, funded by the National Science Foundation.

Michael K. Goodman is a Lecturer in the Department of Geography, King's College London with and PhD from the University of California, Santa Cruz. His interests include the shifting cultural politics of fair trade foods, 'alternative' agro-food networks in the US and UK, and the growth of celebrity environmental politics. Other interests include thinking economies 'otherwise' and critical pedagogical practice in Geography.

Johanna Haas teaches political and environmental geography in the Department of Geography-Geology at Illinois State University. She is in the process of earning her PhD in geography from the Ohio State University, and earned a law degree from the same school. As a West Virginia native, her obsession with mining comes naturally.

Jill Harrison is an Assistant Professor in the Department of Rural Sociology, University of Wisconsin-Madison. Her ongoing research interrogates the structural supports of environmental inequalities through examining political conflicts over agricultural pesticide drift in California. New research focuses on immigration politics and the rise of the Latino immigrant labor force in Wisconsin agriculture. She is particularly interested in community reception of new immigrant populations, changes in the enforcement of immigration policy and border politics.

Peter Hershock is Coordinator of the Asian Studies Development Program at the East-West Center in Honolulu, Hawaii, and holds a doctorate in Asian and comparative philosophy. His research focuses on the relevance of Asian philosophical perspectives in addressing such contemporary issues as: technology and development, education, human rights, and the role of values in cultural and social change. His books include: *Reinventing the Wheel: A Buddhist Response to the Information Age*; *Technology and Cultural Values on the Edge of the Third Millennium*; and *Buddhism in the Public Sphere: Reorienting Global Interdependence*.

Logan A. Hennessy is an Assistant Professor of Social Sciences in the Liberal Studies program at San Francisco State University. He completed his Master's and PhD at UC Berkeley's Department of Environmental Science, Policy, and Management. Logan's interests are in indigenous environmental politics, the environmental history of Latin America and the Caribbean, and the political economy of extractive industries. His other work includes studies of economic transformations and indigenous resistance in Guyana.

Brent McCusker is an Assistant Professor of Geography at West Virginia University. His research focuses on land use and land reform in northern South Africa and land use and livelihood systems in southern Africa. He has written on land use issues in Limpopo, South Africa and continues to conduct research there and in Malawi and Mozambique. In much of his writing, he specifically explores how GIS and political ecology research methods complement and contradict each other, particularly in contentious situations.

Dustin Mulvaney received a PhD from the Department of Environmental Studies at the University of California, Santa Cruz. His work focuses on the anti-genetic engineering social movement and the regulation of three genetically engineered commodities: salmon, maize, and rice. This research reveals the power asymmetries at work in the regulation and adoption of new agricultural technologies and the extent to which anti-genetic engineering activism is shaped by agro-food system restructuring, capacity to enroll the state, and a motivation for broader social change.

Albertus Hadi Pramono is a PhD candidate at the Department of Geography, University of Hawaii at Manoa, Honolulu. Currently his main research interest is on counter-mapping movement with a keen attention on related issues, primarily the interactions between different knowledge systems, indigenous movements, and the history of cartography. His publications centers around this interest, including 'Facing Future: Encouraging Cartographic Literacy in Indigenous Communities' (with Jay Johnson and Renée Louis) published in *ACME: an International E-Journal of Critical Geography*. He has also been actively engaged in environmental movement in Indonesia since 1991.

Karen Schmelzkopf is an associate professor at Monmouth University at the Jersey Shore in New Jersey. She does research in grassroots activism, and has published work on the community gardens in the Lower East Side of New York. It was at one of the gardens that she first heard about Vieques and the Navy occupation. She is currently looking at the role of North Americans in Vieques tourism.

Krisnawati Suryanata is Associate Professor of Geography at the University of Hawaii at Manoa. Her research interests are in agrarian change, resource access, and political ecology with a regional focus on Southeast Asia (especially Indonesia) and Hawaii.

Eve Vogel completed her PhD in the Department of Geography, University of Oregon in the Fall of 2007. She has long been involved in environmental policy, education and politics, from cloud forest management in the Peace Corps in Honduras to teaching environmental education and high school biology in Portland, Oregon. She has been fascinated with the Columbia River and its salmon since moving to Oregon in 1991. Her research interests focus broadly on the history, ideals, institutions, politics and practice of place-based conservation and development efforts. She is working on a book manuscript tentatively entitled *The Regional Power of the Columbia: A Political and Environmental History of the Pacific Northwest's Hydroelectric River*.

Foreword

The 'End' of History (and with it Geography too) is loudly and proudly proclaimed as the inevitable outcome of globalization, the spread of liberal-democracy, and the commodification of everything. Patent nonsense though it is, this line of reasoning is powerful in its suggestion that there is no alternative to 'the' process. As such, it is firmly incumbent on those who would resist this process to reiterate loud and clear for those who seem to be hard of hearing and thinking in these highly distracted times that there are indeed alternatives in thought and practice to such hegemonic discourses.

The present volume aims to do precisely this by zeroing in on the 'contentious' qualities of modern life – qualities that alert us to the cacophony of dissent and alternative ways of thinking and doing, and hence, precisely to those alternative perspectives to the End of History and Geography viewpoint that today's powerbrokers so firmly wish to deny. According to the Oxford English Dictionary, *contentious* means various things: notably, 'argumentative, quarrelsome', 'likely to cause an argument', 'disputed' and 'controversial'. Now, it seems to me, that 'contentious geographies' (and histories, etc) are precisely what is needed today, and the present volume happily serves up bucket loads of contention for the reader's delectation and edification. The reader looking for a happy story of human progress and solidarity (isn't that what the 'End of History and Geography' is supposed to mean?) ought therefore to look away now.

Crisscrossing themes of space, place, scale and environmental knowledge, the book underscores precisely how hegemonic viewpoints are constructed and deployed – be it in popular 'factual' best sellers on the collapse of civilisation or the application of GIS technologies to take but two examples – and how these efforts are challenged at both material and discursive levels. A wide variety of topics are canvassed herein but what unites them is an admirably cussed determination to disagree with conventional mainstream accounts and/or to put in the frame people who are fighting those accounts. As this book underscores, a world of geographies is also and always a world of contentions (and vice versa). As a result, there is too always a world of alternatives from which to choose. In the process of pointing this out, this book ought to serve an invaluable purpose – namely, to be a pebble in the shoes of the powerful.

More than that, though, *Contentious Geographies* forms part of a much wider geography of hope that is keenly needed in these dispiriting times. While this book does not contain all (or even many of) the answers (however, which book could ever do that?), it does contain an awful lot of questions – and that's a good start. Indeed, it is a work whose very contentiousness gives cause for cheer – thereby firmly launching the reader on the path to a more hopeful future. But that is altogether a different contention…

Raymond L. Bryant
Department of Geography, King's College London

Preface

This volume is a manifestation of many seminar and conference discussions, talks and informal conversations over the past seven years. One of the foundations of this project has been the graduate-student organized and facilitated *Political Ecology Working Group* at the University of California, Santa Cruz. Initially advised by Roberto Sanchez-Rodriguez, the group began meeting weekly in 2001 to discuss current work in the field of political ecology. Founding members of the group included Mike Goodman, Max Boykoff, Christopher Bacon, Roseann Cohen, Margaret Fitzsimmons and David Goodman. Over the next five years, the group grew to include consistent involvement from Dustin Mulvaney, Doug Bevington, S. Ravi Rajan, Rebecca Scott, Jill Harrison, Eunice Blavascunas and Anna Zivian, and drew in people from a range of departments such as Environmental Studies, Sociology, Political Science, and Anthropology. Through collaborative development of the reading lists and rotating graduate student coordinators, participants developed a vibrant and constructive space to challenge ideas, perspectives and each other, as well as collectively navigate through the literature and emerging scholarship. (Thanks to Dustin Mulvaney, an archive of course syllabi is lodged at http://envs. ucsc.edu/pewg.)

During this time, Mike Goodman took on the task of organizing six related sessions for the 2004 Association of American Geographers conference in Philadelphia, under the theme *Environmental Contentions: Working through Meanings and Scale*. In addition to many of the authors in this volume, participation in the sessions included the incomparable discussant performances of Nathan Sayre, Lucy Jarosz, Ken McDonald (who stepped in with minutes to spare and did a stupendous job!), Peter Walker, and Francis Harvey and papers given by Kathy McAfee, Ariane de Bremond, Rose Cohen, Joanne Sharp, Julie Sharp, Jennifer Brewer, and Tara Maddock.

The sessions generated much wider discussions that inspired the volume you now hold in your hands. In this, we have viewed these critical inquiries into political ecology a bit like a nice crème brûlée: While sometimes challenging to break its shell, it is filled with rich possibilities and a growing tool-kit from which to explore and conceptualize the often conflict-ridden human-nature nexus. We feel this volume presents a wide range of excellent research that not only begins to crack this shell, but – by creating a space for both early- and established-career researchers – works to extend a broadly-conceived 'political ecology' into new and worthy frontiers.

For Vance, Elijah, Calvin and Augie

Acknowledgements

There are many people we would like to thank for their support at the various stages of this project. First and foremost, we thank Val Rose, Carolyn Court and Aimée Feenan for their tireless perseverance, assistance and guidance in seeing this project to completion. We would especially like to recognize Aimée's editorial efforts and prowess in getting this volume to print. We would like to also thank the original participants in the 2004 AAG sessions that this volume comes out of as well as all the discussants who provided valuable feedback for all of the papers presented. In particular, we want to thank the volume's authors for their contributions but also their fortitude as they had to endure continual deadlines and delays in getting this thing to print; they persisted with much understanding and fortitude.

We want to acknowledge and deeply thank comments on the volume's introduction by Jon Cloke, Raymond Bryant, and Emma Hinton as well as the incredibly close reading and (always) thoughtful suggestions given by the ever-generous David Goodman.

Mike Goodman wants to thank the following colleagues who have shown me nothing but support and valuable feedback in my short time at King's: Raymond Bryant, Michael Redclift, Hannah Cloke, Rob Francis, Mike Raco, Tim Butler, Angela Gurnell and David Demeritt. I'd like to acknowledge and thank the work of Kyle Evered, who was instrumental in the early stages of the project and its completion. In particular, I want to thank Max Boykoff; without his hard work, insight and friendship this book would not have come off as it has. David Goodman has continued to provide me with inestimable intellectual support and guidance throughout this and all my other projects; it has been a continual source of pleasure to work with him over the years. Emma Hinton has quickly become a rigorous 'reality check' and a limitless source of feedback on all sorts of dodgy work; in a very short time, she has proven an invaluable colleague and I thank her for her time and interest (and beer-selection abilities).

The McBride/Blackwood clan was instrumental in smoothing my family's move to the UK and has been supportive in more ways than they can know; a special note of inestimable thanks goes to Susie McBride and Mary Blackwood. Iain Lindsay and the Vanian family have been good friends and we are lucky to know them so well. I want to gratefully acknowledge the past and continuing support of my family from the Whorton tribe of Dave and Lisa Whorton, John Whorton and 'Pa' Whorton; we would not be where we are today with out them. Many heartfelt thanks go to Ken and Joyce Goodman and Sharon Whorton for their continuing help and support. Elaine Goodman put in the groundwork for most of what I do; she is an amazing educator, whose compassion for her students and making the world a better place provides a guiding light for many, but especially for me. Vance's determination and energy – especially the early morning wake-up calls that prompted me to then head back to working on this volume – has proven to be a source of both inestimable joy and inspiration. Mostly though, I must thank and acknowledge the personal help and support of Janet Whorton Goodman, who through all of the trials, tribulations and good times, has been there with invaluable advice, a helping hand and a (poignantly) critical take on things. My deepest thanks and appreciation go to her.

Maxwell T. Boykoff: I would first like to thank Mike Goodman for his tireless work and admirably positive attitude with this book project as well as for his friendship and unyielding enthusiasm for our associated research pursuits. Second, I thank David Goodman, Andrew Schwebel, Michael Loik, Margaret Fitzsimmons, E. Melanie Dupuis, S. Ravi Rajan, Diana Liverman, Emily Boyd, Scott Prudham, J. Timmons Roberts, Sam Randalls, Dave Frame, Emma Tompkins, Ian Curtis, Rose Cohen, Jill Harrison, Dustin Mulvaney, Bryce Sullivan, Alan Richards, Dana Takagi, Mike Urban, David Banghart, Roberto Sanchez-Rodriguez and Angela Y. Davis for each playing a pivotal role in shaping my continuing passions and research pursuits. Third, many friends and family have certainly (often unknowingly) provided perspective and inspiration, such as Monica D.M. Boykoff, Alvin and Marjorie Schoenbeck, Bert and Lois Muhly, Thomas Boykoff, Carl and Leah Moore, Jules Boykoff, Molly Boykoff, Alan Graves, Derek Riebau, Ignacio Fernandez, Jeffrey Betts, Noel Kinder, Colin Ozanne, Kevin Coleman, Erik Ragatz, Andrew Marcin, Jim and George Cadman, Nina Katovic, Elijah and Calvin Boykoff, Roberta Ross Moore, la familias Bustillo, Corrales y Olivera, and Susan Schoenbeck. 'Think, Plan, Study, Play, Rest' said the silent swimmer.

Kyle Evered would like to thank Emine and Augie Evered.

Chapter 1

Contentious Geographies: Environmental Knowledge, Meaning, Scale

Michael K. Goodman, Maxwell T. Boykoff and Kyle T. Evered

In the 'liberal' democratic spaces promised us by Fukayama's *End of History* (1992), something slightly sinister has arisen to capture the attention: the so-called *resource wars* of the twenty-first century (Le Billion 2004). These are not new phenomena as earlier inter- and intra-state conflicts for resources fostered during colonial expansion and de-colonization clearly demonstrate. However, it seems as if this over-determined consumerist moment fuelled by globalization has intensified and entrenched conflicts and violence for a multitude of resources. For example, armed conflicts have exploded for the 'spectacular' commodities of gold, diamonds and – the mother of all commodities – oil, each of which has received its own 'spectacle-ized' Hollywood treatment.[1] At the same time, violence has followed in battles over access to and extraction of the less-than spectacular. The most prescient of these conflicts includes struggles over Congolese coltan (*The Independent*, 2006; see also Cook et al. 2007), a mineral used for the manufacture of Modernity's 'necessary' gadgets of cell phones, computers, and video game consoles.

A number of volumes such as *Green Guerillas* (Collinson 1997) and *Resource Rebels* (Gedicks 2001) have grappled with the temporal and spatial aspects of resource conflicts. Insightful work on these topics (Nevins 2003; Le Billion 2001, 2004, 2007; *Retort*, 2005), particularly the *Violent Environments* collection (Peluso and Watts 2001), has sought to dispense with the simplistic, neo-Malthusian trifecta as the basis for these conflicts – scarcity, poverty, and population growth – to instead offer up more nuanced models of causation steeped in geographical and political economic reasoning. This and allied work in political ecology has contested the equally simplistic assertion that natural environments are merely the stage upon which human actors battle for epistemological and material domination. Rather, from a number of interesting philosophical and empirical angles (e.g. Blaikie 1985; Blaikie and Brookfield 1987; Braun and Castree, 1998; Castree and Braun, 2001; Robbins 2004; Whatmore, 2002), it has been forcefully argued that environments and human societies are *co-constructed*. In this way, for example, violent struggles for African 'blood diamonds' *shape* and are *shaped by* the spatial, ecological, social,

---

1   e.g. *Syriana* (2005) (Middle Eastern oil), *Blood Diamonds* (2006) (African diamonds), *The Rundown* (USA)/*Welcome to the Jungle* (2003) (UK) (Amazonian gold mining).

economic, symbolic, and political characteristics that actively produce these conflicts (Le Billion 2006).[2]

And yet, so much more is afoot today and operating in parallel with the media-grabbing stories of these sorts of 'brute' armed brawls over 'primitive' accumulation and dispossession (cf. Harvey 2003). Important as these may be, this is precisely where *Contentious Geographies* fits in: the volume is concerned with the more 'subtle' contestations over not just resources but the very environments and landscapes that sustain people's *livelihoods*. While controversial geographies have long been manifest through pressing human-environment challenges, this volume explores the active and intimate material, social *and* discursive relationships underlying contemporary yet more diffuse forms of environmental conflict. Thus, these chapters – from conceptual approaches to empirical cases – work to capture the nuanced processes that shape discursive representations and material manifestations of struggles over resources, landscapes and livelihoods.

Under interrogation are the multiple ways that knowledge and epistemic framings shape and are shaped – in short, co-constructed – by environmental conflicts and environments *in the making*, embody *power* through discursive and technological regimes, and translate into political economic and ecological *outcomes*. While the spaces of environmental conflict seemingly generated by the neo-liberal economic imperatives of privatization and profit (Mansfield 2007; see also McCarthy and Prudham 2004) form the backbone of many of the stories being told, the volume looks to explore these contentious geographies in a different register. Namely, the discussions opened up here are about the specific conflict-ridden governance of the environment – and of its contested social and material constructions – through *media, science, policy* and *technology*. Under interrogation, then, is how these particular forms of environmental governance inscribe and perform 'enclosures' of the material *and* discursive kind to bound environments, landscapes and resources for some and open them up for others. Winners and losers, of the natural and human variety, can be created with the stroke of a pen (and now click of a mouse) as much as through the blade of a bulldozer or the barrel of a gun.

Environmental conflict, in all its multifaceted variety – from violent struggle, to political confrontation and direct action, to semiotic and discursive argumentation through media forms – is often produced through *resistance* to particular instances of governance and their deployment through acts of power. In short, in many cases, without some sort of resistance there would be no conflict, a point not lost on many (e.g. Bryant and Bailey 1997; Robbins 2004; Peet and Watts 2004; Neumann 2005) and certainly not the Foucauldians (Rutherford 2007). The stakes of this resistance, though, seem to be bifurcated along global lines: in the South, struggles are most often over *livelihoods* while those in the North are about *lifestyles*. Obviously this formulation is too simplistic since geographers (e.g. Bebbington 2003; Bebbington and Batterbury 2001; Bebbington and Kothari 2006; de Haan 2000; Rocheleau et al. 1997) and others (e.g. Gottlieb 2001; Bullard 2000) tell us it is not at all sensitive to

---

2   For allied and useful takes on theorizing about resources and resource extraction spaces, see Gavin Bridge and colleagues' recent (Bakker and Bridge 2006) and past work (Bridge 2001; Bridge and McManus 2000; Bridge and Jonas 2002).

the shifting scales of environmental and livelihood struggles, especially those based on the precepts of race, class and gender. Yet, what these environmental resistance movements seem to have in common – from California, to Columbia, to Calcutta – is the absolute importance of the *framing* and *re-framing* of environmental struggles in and through *discursive* and *knowledge* practices as part and parcel of this confrontational praxis.

These concerns over the confluences of the 'lived' environment and knowledge frames are what Escobar touches on in his plea for 'discursive materialism' to be placed at the center of what he terms a 'post-structural political ecology' (1996; but cf. Watts and McCarthy 1997). Drawing on his formulation, in environmental struggles, then, words, concept and policy formulations, and their hegemonic deployments act to construct the boundaries of the human-environment relationship to (most often) the benefit of the powerful. In confronting these enclosures, environmental movements very often now battle back on this discursive plane through a range of different semiotic and symbolic formulations. Thus, contentious environmental resistances should be thought of as the multi-form material *and* discursive/semiotic articulations of 'alternative' and/or 'different' histories, livelihoods, worldviews and policies (e.g. Peluso 1992, 1995) that work to act locally, nationally and, more often now, globally.

With all of this in mind, there are three specific aspects of environmental struggles and conflicts that inform the chapters in *Contentious Geographies*; these are *environmental knowledge*, *meaning* and *scale* and we now look to briefly discuss each in turn.

Environmental Knowledge Struggles

Struggles over environmental knowledges and their regimes are at the heart of this volume. From scientific and technological knowledge, to that of the 'Other' in 'indigenous' and local knowledge, how and what knowledge is deployed in environmental management and policy – indeed what *counts* as knowledge let alone *environmental* knowledge – is of paramount consideration. This has been a clear concern of many involved in the social studies of science and knowledge for some time now (Latour 1999, 2004; Forsyth 2003); for our purposes here in thinking about environmental conflicts and knowledge struggles, the focus of this work falls across two axes. The first concern involves the construction of scientific and knowledge networks in the material and discursive, human and 'non-human' artifacts that differentially enable things like science, policy and politics to take shape and become praxis. As a focus for many in the volume, the processes by which the 'hybrid networks' of scientific and Other environmental knowledges come to actually embody this hybridity and produce 'net-workings' are laid bare in the spaces of contentious environments and livelihoods. A second diffuse concern across this work involves the politics and political consequences of network construction; drawing on feminist re-interpretations of network formulations (e.g. Star 1991; see also Whatmore 2002), the focus is on whose knowledge counts and why but also the sorts of politics affected by particular knowledge networks. Many of the chapters, taking an implicit lead from this second theme, are concerned with the politicized processes by which environmental knowledge(s) is/are produced through various media forms and technologies, from popular scientific texts, to news media, policy, law, and novel technologies, and the sorts of political and ecological outcomes that result. In short, this all points to the ways in which science, scientific knowledge and Other knowledges – and, indeed, most importantly, their creation and deployment – are clearly the stuff of politics inside and out.

Conflicts over Environmental Meanings

Meanings invested in everything from landscapes, to 'home', 'place' (e.g. Castree 2003b; Massey 1999), 'science', resources (e.g. Rocheleau and Edmunds 1997), commoditized resources (e.g. McSweeney 2004), and technologies (e.g. Bebbington 2004) are complicit in shaping and engendering environmental struggles. Often, conflicts – violent, political and livelihood – take shape from the constructed and shifting meanings of various environments and their constituent parts. For example, environmental struggles have arisen over the governance of which aspects of 'nature' might be suitable for extraction and commodification and which might not (Katz 1998; Neumann 1998, 2000; Peluso and Watts 2001, *passim*; Robertson 2000; Zerner 2000; *Antipode* 2007). Seeing the environment through the exclusive lens of neo-liberal economism does not always square with different competing worldviews, local or otherwise, although the tropes of 'mainstream' sustainable development and ecological economics are quickly stamping out Other worldviews in the continued name of 'green developmentalism' (Adams 1990; Castree 2003a; McAfee 1999). Through specific case-studies, this volume interrogates how meaning is constructed and manifested through *both* the ontological conditions of nature and the contingent social and political processes involved in architectures that shape the differing interpretations of this nature (Haraway 1997).

Scaled Environmental Struggles

While many have argued that scale is thoroughly socially and politically constructed (Herod 2003; Brenner 2001; Marston 2000; Swyngedouw 1997) nowhere is this more true than in environmental conflicts; the politics of scale is alive and well in environmental struggles, not the least in the 'David versus Goliath' aspects of locals-versus-global-corporation-'X'-theme seemingly repeated at every research or popular-media turn. Thus, in this, we, and many of the subsequent chapters, are not as prepared as some to jettison the concept of scale and its use as an analytical device (Marston et al. 2005). Rather, much to the contrary, *Contentious Geographies* is devoted to exploring how the politics of scale and, in particular, the politics of scale *construction* and *manipulation* are performed by actors from the grassroots, civil society and the state. In particular, understanding how scales are 'jumped' or 'shifted' is integral to assessing the multi-level power dynamics and relations that operate across biophysical and socio-political articulations of local, national, and global political ecologies (Glassman 2001; Swyngedouw 2000, Massey 2001, Bracken and Oughton 2006).

Knowledge, meaning and scale are inextricably intertwined in environmental contentions; below, in the rest of this brief introduction, they form the base of the discussion of the various organizing themes of the volume as well as the core focal point for its chapters.

Translating Contentious Environmental Knowledge and Science

Just beneath the thin yet oft-authoritative veneer of 'crisp' science and eco-political discourses are variegated messy and contentious interactions of knowledge, power and ideology. These contemporary spaces – centred as they are on Western scientific knowledge creation and dissemination – are highly contested, characterized by uncertain facts, politicized interpretations of science, disputed values and intensely debated alternatives (Funtowicz and Ravetz 1990). Indeed, in places where multifarious (environmental) scientific conflicts have arisen, the stakes and tensions of the social, cultural, political, economic and ecological kind are frequently palpable. As asymmetries of scientific 'truth' creation shape and are shaped by uneven social and political terrain, the articulations of these discourses are particularly privileged ways of knowing (Jasanoff 2004) that act on the environment in particular and contentious ways. Over time, dominant discourses and icons of science – and here in particular, environmental science – become entrenched and solidified, and consequently institutional activities and actors get tethered to surrounding storylines that, again, impact landscapes and livelihoods in specific 'good' or 'bad' ways (see Robertson 2006, 2004). Alternatives, in the often differing worldviews and interpretations of 'science' and, indeed, those of the 'dispossessed' of the global South, lurk in the crags of these landscapes; yet, more often than not, the governmentality of the environment is the domain of the privileged and now 'everyday' voices of the entrenched bodies of environmental science, policy and practice fostering, if anything, almost guaranteed contentions over landscapes, resources, and livelihoods (Rutherford 2007).

The contributions in this section call into question and critically explore key processes and effects of the translation and representation of environmental knowledge, science and politics and the subsequent conflicts that ensue. They draw from social constructions of science and critical realist perspectives (e.g. Demeritt 1998; Forsyth 2003) in order to critically analyze this dynamic human-environment interface. In these high-stakes arenas, the chapters explore how perceptions and behaviours are made (in)visible through exegetic perspectives, ideologies and values of what might be called 'environmental science'. Indeed, as mobilized by both tacit and explicit discursive frames, global environmental problems in the forms of (nothing short of) the 'collapse' of global ecological and human civilization, global warming, and water shortages, are manifest locally as well as through collective imaginaries in multiple media and policy formulations. Jasanoff and Wynne elucidate this power of framing in science policy praxis:

> [F]ramings do not flow deterministically from problems fixed by nature, but ... particular framings of environmental problems build upon specific models of agency, causality and

responsibility. These frames in turn are intellectually constraining in that they delimit the universe of further scientific inquiry, political discourse, and possible policy options (1998, 5).

Thus, fighting against these overtly 'positivist' and bounded framings of environmental science (as hegemonically constructed), more democratized approaches to scientific knowledge work to not only muddy the waters but also challenge and shift the composition of these science-practice intersections (Latour 2004). As Beck (1992) outlines in his book *Risk Society*, these other approaches represent a new epoch of challenges that thus demand new conceptualizations and novel academic and popular responses. Each chapter in this section works along this axis of 'thinking' scientific knowledge differently, both conceptually and empirically, and, through case studies, works to cast a critical eye over the spatial and temporal interstices of what counts for environmental science today.

Tim Forsyth, in particular, works through these issues in his analysis of Jared Diamond's bestselling book *Collapse: How Societies Choose to Fail or Survive*. He writes that Diamond takes a detour around many important contested and uneven social, economic and political causes of environmental conflict to instead traverse more ahistorical roads on natural limits causing collapse. As a vehicle through this landscape, Forsyth posits that the book utilized scientific lexicon and authority to give intellectual purchase to advance particular political values and perspectives. Furthermore, through the discursive traction that this best-seller has achieved, Forsyth argues that wider considerations of social vulnerability and development interventions are dangerously subsumed. Consequently, he argues that Diamond then restricts political and policy considerations and debates in a constricting and conservative manner regarding environmental change through such an approach. The chapter makes the claim that greater attention paid to the interactions of social values and environmental knowledge can regain these lost spaces.

Maxwell T. Boykoff examines struggles over the 'semantic drift' of anthropogenic climate change science as it interacts with policy discourse via US mass media. The chapter analyzes comparative as well as historical threads of discourse in order to help explain how US media representational practices have depicted conflict and contentions rather than coherence and convergent views regarding scientific explanations of anthropogenic climate change. Tracing two decades of television and newspaper coverage, Boykoff pays particular attention to various ways in which claims-makers – through asymmetrical power – have influenced media discourses. The chapter analyzes how various truth-claims about the science have gained particular visibility through mass media outlets to thereby contribute to institutional and policy conflicts and considerations at multiple scales. The chapter explores how meaning is maintained and contested through feedbacks of dynamically intertwined socio-political and biophysical processes, and how this case-study informs related environmental conflicts and debates.

Jessica Budds examines how the combined forces of discourse, socio-political and bio-physical power and agency shape conflicts in Chile's La Ligua river basin. This case-study draws upon uneven political and economic pressures as they relate to fruit plantation and small-scale peasant farmer water demands, food security and livelihoods in this region. The chapter analyzes how dynamic and competing factors produce particular assessments of water conflicts, such as conceptions of the physical hydrology of the basin, and notions of water scarcity as well as need. She points to how the framing of solutions as primarily administrative and technical has resulted in problematic socio-ecological outcomes. Furthermore, Budds argues that insufficient attention to nuanced differences such as the heterogeneity of the water users and the spatially-divergent needs of the actors has exacerbated ongoing conflict. Through a 'hydro-social cycle' approach, Budds unpacks and critiques how basin-scale hydrology policy discourse and actions unfold through nature-society interactions, power and agency.

Conflicting and shifting environmental knowledges, livelihoods, and power

Environment and development are inextricable interconnected – and not least in the global South. Equally connected are the lines of power that stitch environment and development together as can be divined in the shifting winds of green developmentalism in institutions like the World Bank. Investigations of these lines of power in development, how they coalesce, their historiography, their broad effects on the conceptualizations and management of (Southern) landscapes, environments and people, were very much a part of the project of development scholars – often referred to as 'post-development' – through out the 1990s and into the new millennium (Escobar 1995, 1998, 2001; Crush 1995). Yet, even with a more 'radical' project of fostering local empowerment and alternative visions of development for 'the people', many felt this work suffered from a lack of theoretical and empirical engagement with the actual *hurly-burly* and *everyday-ness* of development in the shadows of a diverse neo-liberal project (e.g. Watts and Peet 2004). Thus, if anything, spatially-inflected research has become even more relevant to understanding and exploring the *geographical* and *material* specificities and connections of global economic forces to the local, situated (and often deteriorating) livelihoods of the global poor (Blaikie 2000; Hart 2001, 2002, 2004).

In the receding tides of post-development and its penchant for deconstructionism, the analytical figure of 'livelihoods' has arisen to claim the conceptual space for understanding development and environments as lived, practiced, and imagined. Drawing on Bebbington's work (2000, 498), this materialist corrective works to understand 'how people make a living and [how they] mak[e] it meaningful'. As he continues,

> [m]ore viable livelihoods will not be romanced into existence, but must instead be built up from already existing, and however imperfect strategies. Understanding livelihood thus becomes critical for theory, in order to understand how places are produced and governed, and who participates in these processes. It is also critical for practice – to understand the ways in which people have created livelihood opportunities that foster accumulation as well as the obstacles to such accumulation (Bebbington 2003, 515).

Thus, 'building livelihoods' often means, in this age of structural adjustment and neo-liberal projects, conflict and contention – environmental or otherwise – with current political economic power structures at local, regional and global scales.

And part of this task, thusly, might just mean building 'alternative' livelihoods that include everything from 'transnational' (Bebbington and Kothari 2006; Bebbington and Batterbury 2001) and 'diverse' economic networks (Gibson-Graham 2005a, 2005b),[3] to novel forms of trade (Goodman 2004) and Other identities and knowledges (Perrault 2001, 2003a, 2003b). Key, with respect to (alternative) livelihoods, is one of Geography's most sacred concepts: that of 'place'. Indeed, as Bebbington (2003: 302) puts it,

> place and livelihood clearly intersect as, to a considerable extent, places are produced out of livelihoods of people, while at the same time structuring elements of those livelihoods. But clearly neither livelihood nor place is ring fenced. Thus any discussion of place and livelihood must also be infused with concerns for *scale* and *networks* (original emphasis).

Here, then, livelihoods are always contentious, especially as they are built *in place* but also *out of place* through the myriad of connections that work to construct and, indeed, conflict with livelihoods and places.

Trevor L. Birkenholtz – through the lens of Agarwal's (2005) 'environmentality' – explores the contentious deployments of groundwater conservation regulation in Rajasthan, India. Here, he sees a fundamental rift in local and state groundwater knowledge that will make the 'roll-out' of these regulations uneasy at best and potentially disastrous (and doomed to failure) at worst in the ultimate goal of water conservation and livelihood sustainability. Framed more specifically through Agrawal's concepts of 'governmentalized localities' and 'environmental subjects', the crux of the problem springs from the different legitimized knowledges of groundwater management that circulate in local and regional management networks. In short, for a number of interesting reasons related to trust, praxis and cost, local farmers would very much rather consult *Sungha* – Hindu water diviners – and tubewell drilling firms than government engineers in locating and operating their wells. Top-down government regulation, even in the context of physical violence, has failed to grasp the importance of the 'Other' viable knowledges provided by both the *Sungha* and tubewell firms; State groundwater institutions are quite simply running rough-shod over the practices of the local farmers whose trust, Birkenholtz argues, must be gained and knowledge valued in order to promote groundwater conservation. He suggests that shifting the balance of power more into the hands of local farmers and utilizing local knowledge about groundwater management might work to develop a much more 'equal' and successful water conservation program in Rajasthan.

Logan A. Hennessey's chapter conducts a fascinating analysis of the various discourses that have sought to 'fix' the various spatial and historical identities of the Huaorani in Ecuador, not surprisingly, to the detriment of their own self-determined livelihoods. In an engagement with the literature around indigenous identities and the ties this has to the 'defense' of livelihoods (e.g. Escobar 2001; Castree 2004), Hennessey examines how various discourses of the Huaorani generated from the

---

3   But see Kelly (2005), Laurie (2005) and Aguilar (2005) for a critique of Gibson-Graham's return to some of the blind spots of post-developmentalism.

'outside' by others – conservation NGOs, the State, and oil firms – attempt to construct these people and their environments to the distinct advantage of powerful extra-local players and institutions. Indeed, even the 'good guys' of the Rainforest Action Network (RAN) are not innocent; here, he argues, RAN papers-over the diversity of self-constructed livelihoods that make up the 'modern' incarnations of the Huaorani, to instead construct a homogenizing discourse of the 'noble savage' and equally savage and worthy tropical nature worth saving. Thus, the contentious political and environmental projects of these outside actors – either conservation and/or oil extraction – gain traction in the contradictions of these 'fixing' yet 'elastic' discourses; it is the multiplicity and content of these discursive 'spearpoints' that works to disempower and disassemble the Huaorani's ability to determine their own development in a more cohesive manner.

Environmental movements: Contested (re)scaling of knowledges, problems, and narratives

At the core of their existence, environmental movements are contentious entities. They exist only in so far as they work to challenge privileged knowledge, meanings, and scales – ostensibly on behalf of the 'public' in total or for some other environment-related 'cause'. All of this, though, it must be remembered, is mobilized through particular notions and mobilizations of 'nature' and 'environment'. In certain cases, mobilization of such interests seeks structural re-definitions or reforms, while in other situations, movements may resist outright existing knowledge-power assemblages with re-placement or revolution being the actual goal. Thus, regardless of its intended outcomes – to attempt to re-structure or re-invent existing modes of environmental governmentality – environmental movements are greatly concerned with mobilizing real and/or discursive refutations of existing and/or would-be hegemonies. In the absence of such 'contentious' agendas, they are little more than environmental clubs or advocacy, interest, or lobby groups.

In scholarship on environmental movements (e.g. Peet and Watts 2004), they are often characterized as one of several major types of so-called 'New Social Movements' (NSMs) – along with feminist, indigenous, and other movement groups. While the 'new'-ness of NSMs might be regarded as a profound overstatement (Calhoun 1995), they are one of the most vociferous sources of alternative environmental discourse (Wolford 2005; see also Wright and Wolford 2003; Mauch et al. 2006). Moreover, their varied identities, spatialities, and intellectual, organizational, and programmatic structures reveal not only profound diversity but also wide-ranging linkages with other interests and agendas that may also be advocating for the rights of, for example, particular groups, beliefs, and/or ideologies. This diversity and variation is not only apparent when examining environmental movements but is also most apparent amid contemporary explorations of both their common grounds and cleavages with indigenous, feminist, labor, and other movements. This is particularly true as (some) environmental movements could be said to collectively constitute actors in a wider, transnational 'anti-globalization' – or, perhaps more appropriately, anti-neoliberal – movement (Bandy and Smith 2005; Curran 2006; Leitner et al. 2007; Munch 2007; Reitan 2007).

While environmental movements are often regarded as one of many elements of 'civil society' – a term with profound discursive currency in state-ist and neo-liberal rhetoric of democratization and capacity- and state-building (e.g. Jessop 2002; see also McCarthy 2005, 2006) – not all expressions of environmental civil society are environmental movements. Indeed, though civil society is often defined as encompassing a sphere distinct from state and commercial interests and authority, as growing critical scholarship on Non-Governmental Organizations (NGOs) (e.g. Bryant 2005; West 2006; Elyachar 2005) establishes, many NGOs are not so 'non-governmental' as they might seem.[4]

And yet, despite a drive to institutionalization and the mainstreaming of environmental movements, their contentious capacities to challenge state and corporate governmentality – to contest dominant paradigms and epistemologies – and to find alternative means of fostering social and moral capital are essential 'virtues' amid contemporary environmental struggles. Here we want to flag the sort of 'possibilism' (Gibson-Graham 2005a) that might be present in environmental movements to carve out spaces of action and, indeed, even political change (Routledge 2003). One of the key ongoing 'contributions' of these movements has been the 'nitty-gritty' work they have done by putting forward alternative knowledge claims about the environment, how to 'see' it, and what it might be for. Here, the power of environmental movements has been in their ability to make spatial and discursive connections between the global and the local and the local and the global when conceptualizing environmental problems and solutions. This has not only resulted in the growing 'mass-ification' of environmental movements themselves and their punctuated events – think here of the 'battle in Seattle' (see Smith 2002; Khagram et al. 2002), the G8 'actions' in Edinburgh (Routledge 2005), and the growing influence of the World Social Forum and its green tinges – but also the abilities of environmental movements to match the global nature of capital and the reach of many corporate institutions. Thus scaling up, out, and down for environmental movements has been important for not only making the movement 'work' at the pace of global environmental issues, but has been instrumental in shaping the discourses around global and local environmental problems and solutions more specifically.

Jill Harrison describes the contentious framings of pesticide drift in California's farmworker communities and the implications this has for workers' and their families' health. The social and material constructions of scale – of the terms and scope of the 'everyday' experiences of pesticide drift versus its regulation and reporting – define this story of environmental and livelihood conflict. Seen through the lens of work on the politics of scale, Harrison details how the framings of pesticide drift and normal 'accidents' (i.e. poisonings) by the State regulatory agencies makes the so-called 'dispensable' farmworkers and their health even more invisible than they already are. Thus, pesticide drift incidents become just that, discrete and isolated incidents rather than the everyday, albeit 'officially' invisible, business of California's industrial agriculture landscape. To counter these absences, activists, as Harrison puts it, 'manipulate' scale temporally and spatially by two principal means. First,

---

4   Much of this can be seen in the growing whimsical plethora of oxymoronic acronyms such as INGOs (international NGOs), BINGOs (business-oriented international NGOs), GONGOs (government-operated NGOs), and QUANGOs (quasi-autonomous NGOs).

they do this by fleshing out the local to provide a sort of thick description about what it means to live and work in farmerworker communities with pesticide drift on a continual basis. Second, activists shift the discursive scale of analysis 'upward' to re-conceptualize pesticide drift as a broader and more indiscriminate problem of air pollution. In the end, it is the State's dominant framing of pesticide drift as local and isolated – a politics of scale if ever there was one – that problematically (re)inscribes the social, health and environmental inequalities present in California and other farming-scapes.

Karen Schmelzkopf paints a geo-historical account of the fight for land and life on the small island of Vieques, Puerto Rico – one of the principal military depots and practice bombing sites for the US Navy for the last 60 years. She describes how activists, after the horrible death of one of the island's inhabitants from a misguided bomb, were able to 'jump scale' and develop strong ties with national and international activists and social movements. Through the connections between the local direct action 'Camp for Peace and Justice', television, movie and music celebrities, and the US Hispanic community, the movement gained enough political power to push the Navy out, an enormous achievement in the 9/11 era of just about everything being trumped by 'national security'. But neither the story nor the struggle ends there; in fact they both might just be beginning. From the Navy presence and constant bombing runs, Vieques is left with a distinctly toxic and indeed, explosive, legacy of unexploded bombs, heavy metal contamination, increasing cancer rates and destroyed landscapes and has now been designated a Superfund site. And, even with all this pollution – to add livelihood insult to environmental injury – much of the land has been designated a National Wildlife Refuge; thus, much of the island has changed hands from the Navy to the Department of the Interior and so, is still very much under the control of the original 'colonizers'. She concludes with a sober assessment of the ability of Viequenses to determine their own livelihoods let alone their own landscapes in light of the continuing powerful and contentious multi-scalar reach of the US government.

Dustin Mulvaney explores the scaling and re-scaling of contentious environmental politics in the development of 'GE-free' zones in two counties in California. While deploying a politics of scale approach similar to Harrison and Schmelzkopf, Mulvaney argues we need additional lenses to flesh out the story of the success of anti-genetic engineering activists in Mendocino County versus their failure in Butte County. Here he underscores that to understand the differential results of these campaigns, we must also understand the politics of place and the discursive storylines (Hajer 1995) that construct, inflect and reflect these material politics of scale, space *and* place. In short, the successful and politicized constructions of who *and what* is an 'outsider' versus 'insider' – from corporations and activists, to money, pests, markets and genes – are what produce the opposite reactions to the GE-free legislation in the two counties. For example, in Mendocino, the activists' campaign was able to coalesce, to a large degree, around a local distaste for 'outside' control by agri-business and corporations. In Butte County the current 'bogey-man' of the 'know-nothing', latte-sipping San Francisco liberal was deployed to garner local opposition to these 'outside' activists' supposed influence in the GE-free campaign. Mulvaney leaves us with a set of important questions about the spatially-inflected political (and indeed, material) tactics of both activists and the biotechnology industry to 'deal' with genetically engineered organisms: namely, questions remain not only about the ecological risk of GEOs but also the discursive strategies used to 'contain' GEOs and their protests.

Contested production of environmental histories, law, and knowledge

In explorations of history and historiography, it has generally come to be recognized by critical scholars as a given that much of what constituted history in Western – and even in non-Western – contexts is reflective of power (Trouillot 1995). Indeed, an examination of simply the membership rosters of formal historical societies in the West makes this abundantly clear.[5] The multi-disciplinary field of environmental history, however, might be seen as having a rather separate genealogy – one that originates well beyond the confines of the discipline of history. Indeed, the contributions of historians like Worster (1979, 1985), Merchant (1980) and Cronon (1983, 1991)[6] stand out today as fine examples of critical scholarship that confronted matters like the exploitation of indigenous peoples and nature, gender, empire, the abuses of capitalist and socialist systems alike, and the implications of varied constructs of nature/environment. Yet, the traditions of historical analysis in the field of cultural ecology – the early grandparent of political ecology (see Robbins 2004; Watts 2000) – reflected a contentious orientation many decades prior to the rise of critical environmental history within history itself. Perhaps most noteworthy in this tradition of authoring contentious histories are some of the works of Carl Sauer and his students.

Although lacking the analytical tools of subsequent neo-Marxian and postmodernist scholars, their works were profoundly 'contentious' in the approach of listening to indigenous community voices and observing their pre-historic records, critically assessing the socio-cultural and environmental consequences of development, and communicating these lessons to the West.This tradition, coupled with later materialist and discursive approaches to environmental history (Demeritt 1994; Williams 1994), has resulted in a contemporary field that is analytically rich and diverse and cognizant of the multiple contentious interpretations of environmental history and of the conflictual legacies of the human-environment relationship in such spheres as environmental policy making.

Just as historical narratives of the environment have been re-cast and forced beyond any singular sort of interpretation through contentious re-articulations of the past, so too do laws and legal systems constitute artifacts and arenas of authority and official knowledge of ecologies that may be re-written and resisted and/or re-

---

5   In the case of the United States, for example, the list of past presidents of the American Historical Association includes advocates of American empire – whether proudly stated or simply functional – like Alfred Thayer Mahan, Theodore Roosevelt, and Woodrow Wilson.

6   For other important texts on this, see also Crosby (1972, 1986), Merchant (1989) and Weiner (1988, 1999).

invented through popular pressure. The scales of contestation of laws and legal systems range from the local to the international – often evidenced in increasing global linkages and effects. In many instances we may envision the typical conflicts over matters of resources, as with struggles for agrarian policy reform (e.g. Wolford 2005), at scales between individuals and/or communities and the state. Increasingly, however, we are also witnessing legal authority as exercised largely by corporate interests coming into conflict with traditional ecologies, marginalized communities, and/or even entire regions over matters of intellectual property and the rights to biodiversity (McAfee 2003), ownership of previously communal or public assets (Spronk and Webber 2007), and matters of access to markets and resources (Kuecker 2007). Moreover, regional environmental conflict is becoming apparent between the so-called South and the North over matters of global environmental governance (e.g. Najam 2005). It is here that environmental regulation through policy is at the heart of drawing the boundaries of what is possible in environmental management, making regulatory schemes one of the most powerful determinants of the human-environment relationship and environmental conflict.

Eve Vogel, in her chapter, examines how the privileging of past regional-scale planning within the United States has fostered environmental contentions up to today. She explores how the environmental regulation of the Columbia and Snake rivers has shaped prioritization and perceptions of management in the Pacific Northwest. Through this case-study that tracks the development of the 2000 decision not to breach the lower Snake river dams, this analysis demonstrates that power has been effectively leveraged and mobilized by the Bonneville Power Administration and allies through particular 'geographical framings' and 'fixing' of scale. The chapter assesses how this primacy of the regional scale has consequently impacted ongoing decision-making as well as imaginaries in the region over the last six decades. By unpacking and historicizing key political economic, social and ecological processes that have led to this institutionalization of space, Vogel strives to uncover how scalar choices shape geographical meaning. Thus, she illustrates dynamic contestations between environmental law and meaning through an examination of how river breaching decisions constrain policy choices. Moreover, through exploring these interactions in the Pacific Northwest, the chapter explains how such decisions contribute to problems that threaten the sustainability of long-term human-environment interactions.

Johanna Haas addresses how federal mining legislation has undergone uneven development and implementation in the Eastern and Western regions of the United States. She attributes this to not only different regional landscape features, but rather, as primarily a function of the regionally-distinct imaginaries and material realities of mining in the East versus the West. With a focus on the 1977 Surface Mining Control and Reclamation Act, Haas places legal trajectories and their contentious environmental effects in historical context. This analysis accounts for how power and scalar elements have contributed to spatially differentiated extractive industrial mining operations. She argues that in each region, divergent views of how legal architectures met human-environment interactions have cast Western companies as caretaker organizations and Eastern groups as destructive entities. These constructions have fed back into functionally different implementation of regulations whereby Western groups have been held to flexible results-based standards and Eastern companies have been forced to follow more inflexible performance-based standards. Through analyses of how mining institutions and laws are constructed and maintained, Haas sheds light on the intertwined and multi-directional processes of environmental law, history and discourse as well as their inseparable material outcomes on the US environment and landscape.

Fraught spatial technologies and knowledge construction

With Geography's history of mapping empire – not to mention the associated justifications *for* empire – the discipline is no stranger to skepticism by critical scholars concerned with matters of spatial information and governmentality. In light of this history of the discipline, and a very similar one within anthropology, the appearance of critical scholarship on novel mapping technologies such as Geographic Information Systems/Sciences (GIS/GISc) should come as little surprise. Moreover, given the over-riding processes and claims to new powerful forms of 'representation' by the field of GIS, it constituted an ideal (and quite justified) target for the deconstructionist scholars of the early- and mid-1990s. Foremost among these critical works was Pickles' edited volume *Ground Truth* (1995), a work with contributions that engaged the particular and potential pitfalls of still-emergent technologies for gathering, managing, manipulating, and representing spatial information and 'data'. Through this work and others (Robbins 2001; Sheppard 2005; Schuurman 2006), came the development of *critical GIS* that sought to challenge the field and emplace accountability among practitioners, end-users, and power-holders alike. Focused on critiquing the underlying theoretical, ethical and practical framings of the field (e.g. Fox et al. 2003, 2005), critical GIS is on the frontline of the 'appropriate' use of this technology and the spatial information it generates as well as the sorts of conflicts and powerful implications the deployments of GIS might engender. Indeed, in some ways critical GIS is about making environments and landscapes *more* contentious by working towards participatory and 'counter' mapping (e.g. Peluso 1995) with usually disenfranchised locals as a way to confront the problems of representation and promote new forms of empowerment.

And yet, unsurprisingly, even with critical GIS's penchant for an analytical focus on methodology, ethics, and the politics of its technological feats, many questions remain. For example, will critical GIS – and, equally as importantly, GIS/GISc as a field – actually foster more than an over-riding obsession with collecting, storing, and representing knowledges about those communities and environments that are disempowered? In short, can participation get beyond 'consultation' of what the 'people' want and the performance of focus groups to 'share' the results of a mapping project to instead include a deeper sense of participation that might build bridges between groups that have generally been separated by socio-economic, political, education, and other cleavages? If critical GIS can be practiced as not just a form of academic critique but rather a contentious *system* of knowledge production and use, it might just make a significant contribution towards both overcoming the power divides associated with 'expert' and scientific knowledge systems (Forsyth 2003, Mitchell 2002).

In their chapter, *Fox, Suryanata, Hershock* and *Pramono* explore how multi-scale uses of spatial information technologies are value-laden, and therefore can influence community-based management in multiple and conflicting ways. Through examinations of case-studies in Cambodia and Indonesia, they examine a set of wider ethical and social questions regarding the uptake of spatial information technologies and related 'ironic' effects. The chapter looks at how the incorporation of these tools can inadvertently bound 'ways of knowing', such as ways of viewing property rights. Therefore, these actions entrench particular values and socio-ethical constructs for both the subjects and objects under study. The authors point out that these context- and scalar-dependent processes can generate conflicting effects, and the chapter draws on three inter-related dimensions where such effects raise ethical concerns. As an illustration of these interactions, Fox et al. describe how mapping can improve *insitu* understanding of community resources and ecosystem services, but also how it can also enhance *exsitu* knowledge for potentially extractive and exploitative activities. By interrogating the uses of these technologies in participatory mapping through wider social, ethical, and cultural lenses, the chapter addresses issues of (dis)empowerment (dis)enrollment and (dis)association in community and environmental mapping.

Brent McCusker examines two spaces in northern South Africa to look to how GIS and remote sensing technologies intervene in multi-scale discourses over post-apartheid property rights and claims. Situating the conflicts in both political and ecological contexts over time, McCusker traces land-rights and land-reform narratives on six farm associations in the Mahlambandlovu and Rondebosch Communal Property Associations. The chapter examines local versus expert knowledge issues and expressions of competing land claims. It then looks at how government officials and power-brokers seek to employ spatial technologies in order to 'settle' debates and discussions. However, considerations of social, ethical and cultural aspects of the issues demonstrate that these interventions are not at all neutral. Overall, this chapter addresses how local knowledges and contentions might be smoothed out through such value-laden technological interventions in order to scale-up and contribute to land reform policies and agendas. McCusker draws attention to how these particular ways of seeing spaces and places have contributed in multiple ways to understanding as well as quickening but also slowing the pace of (dis)possession and land reform in South Africa.

Conclusion: Towards Understanding Environmental Contentions

The contemporary and broad range of environmental conflicts critically examined in this volume – from contested representations of environmental science, to violent displacement, the scaling and re-scaling of environmental knowledge, protest, and regulation, to the contentious technological mapping of people and places – are united in an attempt to locate power and governance on the terrain of nature and society interactions. As the different chapters and case-studies show, in a variety of geographic and political-economic circumstances, environments become and remain contentious in often subtle ways that can threaten local and global ecological landscapes and resources, and most intimately, people's very livelihoods.

Importantly, the volume works to understand underlying discursive conditions and states of knowledge as well as forms of articulation that frame how people perceive and experience current resource and environmental conflicts. It is our hope that in clarifying the powerful and power-laden processes by which environmental knowledge is generated, framed, communicated and interpreted, this volume helps to reveal how environmental conflicts are and can be reconsidered and thus (re)opened to enhance negotiation in efforts to create more sustainable and just futures.

References

Adams, W. (1990), *Green Development: Environment and Sustainability in the Third World* (London: Routledge).

Agarwal, A. (2005), *Environmentality: Technologies of Government and the Making of Subjects* (Durham: Duke University Press).

Aguilar, F. (2005), 'Excess Possibilities?: Ethics, Populism, and Community Economy. A Commentary on J.K. Gibson-Graham's "Surplus Possibilities: Postdevelopment and Community Economies"', *Singapore Journal of Tropical Geography* 26: 1, 27–31.

Antipode (2007), 'Special Issue: Privatization: Property and the Remaking of Nature-Society Relations', 39.

Bakker, K. and Bridge, G. (2006), 'Material Worlds? Resource Geographies and the "Matter of Nature"', *Progress in Human Geography* 30: 1, 5–27.

Bandy, J. and Smith, J. (eds) (2005), *Coalitions across Borders: Transnational Protest and the Neoliberal Order* (Lanham: Rowman and Littlefield).

Bebbington, A. (2000), 'Reencountering Development: Livelihood Transitions and Place Transformations in the Andes', *Annals of the Association of American Geographers* 90: 3, 495–520.

Bebbington, A. (2003), 'Global Networks and Local Developments: Agendas for Development Geography', *Tijdschrift voor Economische en Sociale Geografie* 94: 3, 297–309.

Bebbington, A. (2004), 'Movements and Modernizations, Markets and Municipalities: Indigenous Federations in Rural Ecuador', in Peet, R. and Watts, M. (eds), *Liberation Ecologies: Environment, Development, Social Movements*, 2nd edition (London: Routledge), 394–422.

Bebbington, A. and Batterbury, S. (2001), 'Transnational Livelihoods and Landscapes: Political Ecologies of Globalization', *Ecumene* 8: 4, 369–380.

Bebbington, A. and Kothari, U. (2006), 'Transnational Development Networks', *Environment and Planning A* 38, 849–866.

Beck, U. (1992), *Risk Society: Towards a New Modernity* (London: Sage).

Blaikie, P. (1985), *The Political Ecology of Soil Erosion in Developing Countries* (London: Longman Scientific and Technical).

Blaikie, P. (2000), 'Development, Post-, Anti-, and Populist: A Critical Review', *Environment and Planning A* 32: 6, 951–1140.

Blaikie, P. and Brookfield, H. (1987), *Land Degradation and Society* (London: Methuen).

Bracken, L. and Oughton, E. (2006), '"What Do You Mean?": The Importance of Language in Developing Interdisciplinary Research', *Transactions of the Institute of British Geographers* 31: 3, 371–382.

Braun, B. and Castree, N. (eds) (1998), *Remaking Reality: Nature at the End of the Millennium* (London: Routledge).

Brenner, N. (2001), 'The Limits to Scale? Methodological Reflections on Scalar Structuration', *Progress in Human Geography* 25: 4, 591–614.

Bridge, G. (2001), 'Resource Triumphalism: Postindustrial Narratives of Primary Commodity Production', *Environment and Planning A* 33, 2149–2173.

Bridge, G. and Jonas, A. (2002), 'Governing Nature: The Reregulation of Resource Access, Production, and Consumption', *Environment and Planning A* 34, 756–766.

Bridge, G. and McManus, P. (2000), 'Sticks and Stones: Environmental Narratives and Discursive Regulation in the Forestry and Mining Sectors', *Antipode* 32: 1, 10–47.

Bryant, R. (2005), *Nongovernmental Organizations in Environmental Struggles: Politics and the Making of Moral Capital in the Philippines* (New Haven: Yale University Press).

Bryant, R. and Bailey, S. (1997), *Third World Political Ecology* (London: Routledge).

Bullard, R. (2000), *Dumping in Dixie: Race, Class and Environmental Quality*, 3rd edition (Boulder: Westview Press).

Calhoun, C. (1995), '"New Social Movements" of the Early Nineteenth Century', in Traugott (ed.), *Repertoires and Cycles of Collective Action* (Durham: Duke University Press), 173–216.

Castree, N. (2003a), 'Green Development: Saving Nature by Selling It?' *Geography Review* 16: 3, 12–14.

Castree, N. (2003b), 'Place: Connections and Boundaries in an Interdependent World', in Holloway, S., Rice, S. and Valentine, G. (eds), *Key Concepts in Geography* (London: Sage), 165–186.

Castree, N. (2004), 'Differential Geographies: Place, Indigenous Rights and "Local" Resources', *Political Geography* 23: 2, 133–167.

Castree, N. and Braun, B. (eds) (2001), *Social Nature: Theory, Practice, and Politics* (London: Blackwell).

Collinson, H. (ed.) (1997), *Green Guerrillas: Environmental Conflict and Initiative in Latin America and the Caribbean* (Montreal: Black Rose Books).

Cook et al. (2007), '"It's More Than Just What It Is": Defetishizing Commodities, Changing Pedagogies, Situating Geographies', *Geoforum* 38, 1113–1126.

Cronon, W. (1983), *Changes in the Land: Indians, Colonists, and the Ecology of New England* (New York: Hill and Wang).

Cronon, W. (1991), *Nature's Metropolis: Chicago and the Great West* (New York: W.W. Norton).

Crosby, A. (1972), *The Columbian Exchange: Biological and Cultural Consequences of 1492* (Westport: Greenwood).

Crosby, A. (1986), *Ecological Imperialism: The Biological Expansion of Europe, 900–1900* (Cambridge: Cambridge University Press).

Crush, J. (ed.) (1995), *The Power of Development* (London: Routledge).

Curran, G. (2006), *21st Century Dissent: Anarchism, Anti-Globalization, and Environmentalism* (Basingstoke: Palgrave Macmillan).

de Haan, L. (2000), 'Globalization, Localization, and Sustainable Livelihood', *Sociologia Ruralis* 40: 3, 339–364.

Demeritt, D. (1994), 'Ecology, Objectivity, and Critique in Writings on Nature and Human Societies', *Journal of Historical Materialism* 20: 1, 22–37.

Demeritt, D. (1998), 'Science, Social Constructivism and Nature', in Braun, B. and Castree, N. (eds), *Remaking Reality: Nature at the Millennium* (London: Routledge), 173–193.

Elyachar, J. (2005), *Markets of Dispossession: Ngos, Economic Development, and the State in Cairo* (Durham: Duke University Press).

Escobar, A. (1995), *Encountering Development: The Making and Unmaking of the Third World* (Princeton: Princeton University Press).

Escobar, A. (1996), 'Constructing Nature: Elements for a Poststructural Political Ecology', in Peet, R. and Watts, M. (eds), *Liberation Ecologies: Environment, Development, Social Movements* (London: Routledge), 46–68.

Escobar, A. (1998), 'Whose Knowledge, Whose Nature?: Biodiversity, Conservation, and the Political Ecology of Social Movements', *Journal of Political Ecology* 5, 53–82.

Escobar, A. (2001), 'Culture Sits in Places: Reflections on Globalism and Subaltern Strategies of Localization', *Political Geography* 20, 139–174.

Forsyth, T. (2003), *Critical Political Ecology: The Politics of Environmental Science* (London: Routledge).

Fox, J., Rindfuss, R., Walsh, S. and Mishra, V. (eds) (2003), *People and the Environment: Approaches for Linking Household and Community Surveys to Remote Sensing and GIS* (Boston: Kluwer).

Fox, J., Suryanata, K. and Hershock, P. (eds) (2005), *Mapping Communities: Ethnics, Values, Practice* (Honolulu: East-West Center).

Fukuyama, F. (1992), *The End of History and the Last Man* (New York: Free Press).

Funtowicz, S. and Ravetz, J. (1990), *Uncertainty and Quality in Science for Policy* (Amsterdam: Kluwer).

Gedicks, A. (2001), *Resource Rebels: Native Challenges to Mining and Oil Corporations* (Cambridge, MA: South End Press).

Gibson-Graham, J.K. (2005a), 'Surplus Possibilities: Postdevelopment and Community Economies', *Singapore Journal of Tropical Geography* 26: 1, 4–26.

Gibson-Graham, J.K. (2005b), 'Traversing the Fantasy of Sufficiency', *Singapore Journal of Tropical Geography* 26: 2, 119–126.

Glassman, J. (2001), 'From Seattle (and Ubon) to Bangkok: The Scales of Resistance to Corporate Globalization', *Environment and Planning D: Society and Space* 19, 513–533.

Goodman, M. (2004), 'Reading Fair Trade: Political Ecological Imaginary and the Moral Economy of Fair Trade Foods', *Political Geography* 23: 7, 891–915.

Gottlieb, R. (2001), *Enviromentalism Unbound: Exploring New Pathways for Change* (Cambridge: MIT press).

Hajer, M. (1995), *The Politics of Environmental Discourse: Ecological Modernization and the Policy Process* (Oxford: Clarendon Press).

Haraway, D. (1997), *Modest_Witness@Second_Millennium.Femaleman_Meets_Oncomousetmfeminism and Technoscience* (London: Routledge).

Hart, G. (2001), 'Development Critiques in the 1990s: *Culs-De-Sac* and Promising Paths', *Progress in Human Geography* 25: 4, 649–658.

Hart, G. (2002), 'Geography and Development: Development/s Beyond Neoliberalism? Power, Culture, Political Economy', *Progress in Human Geography* 26: 2, 812–822.

Hart, G. (2004), 'Geography and Development: Critical Ethnographies', *Progress in Human Geography* 28: 1, 91–100.

Harvey, D. (2003), *The New Imperialism* (Oxford: OUP).

Herod, A. (2003), 'Scale: The Local and the Global', in Holloway, S., Rice, S. and Valentine, G. (eds), *Key Concepts in Geography* (London: Sage), 229–247.

Jasanoff, S. (2004), *States of Knowledge: The Co-Production of Science and the Social Order* (London: Routledge).

Jasanoff, S. and Wynne, B. (1998), 'Science and Decision Making', in Raynor, S. and Malone, E. (eds), *Human Choice and Climate Change: The Societal Framework* (Columbus: Battelle Press).

Jessop, B. (2002), 'Liberalism, Neoliberalism, and Urban Governance: A State-Theoretical Perspective', *Antipode* 34, 452–472.

Katz, C. (1998), 'Whose Nature, Whose Culture?: Private Production Spaces and the "Preservation" of Nature', in Braun, B. and Castree, N. (eds), *Remaking Reality: Nature at the Millennium* (London: Routledge), 46–63.

Kelly, P. (2005), 'Scale, Power and the Limits to Possibilities. A Commentary on J.K. Gibson-Graham's "Surplus Possibilities: Postdevelopment and Community Economies"', *Singapore Journal of Tropical Geography* 26: 1, 39–43.

Khagram, S., Riker, J. and Sikkink, K. (2002), 'From Santiago to Seattle: Transnational Advocacy Groups Restructuring World Politics', in Khagram, S., Riker, J. and Sikkink, K. (eds), *Restructuring World Politics: Transnational Social Movements, Networks, and Norms* (Minneapolis: University of Minnesota), 3–23.

Kuecker, G. (2007), 'Fighting for the Forests: Grassroots Resistance to Mining in Northern Ecuador', *Latin American Perspectives* 34: 2, 94–107.

Latour, B. (1999), *Pandora's Hope: Essays on the Reality of Science Studies* (Cambridge: Harvard University Press).

Latour, B. (2004), *Politics of Nature: How to Bring the Sciences into Democracy* (Cambridge: Harvard University Press).

Laurie, N. (2005), 'Putting the Messiness Back In: Towards a Geography of Development as Creativity. A Commentary on J.K. Gibson-Graham's "Surplus Possibilities: Postdevelopment and Community Economies"', *Singapore Journal of Tropical Geography* 26: 1, 32–35.

Le Billion, P. (2001), 'The Political Ecology of War: Natural Resources and Armed Conflicts', *Political Geography* 29: 561–584.

Le Billion, P. (2004), 'The Geopolitics of 'Resource Wars'', *Geopolitics* 9: 1, 1–24.

Le Billion, P. (2006), 'Fatal Transactions: Conflict Diamonds and the (Anti)Terrorist Consumer', *Antipode* 38, 778–801.

Le Billion, P. (2007), 'Geographies of War: Perspectives on "Resource Wars"', *Geography Compass* 1, 1–20.

Leitner, H., Peck, J. and Sheppard, E. (eds) (2007), *Contesting Neo-Liberalism: Urban Frontiers* (New York: Guilford Press).

Mansfield, B. (2007), 'Privatization: Property and the Remaking of Nature–Society Relations. Introduction to the Special Issue', *Antipode* 39, 393–405.

Marston, S. (2000), 'The Social Construction of Scale', *Progress in Human Geography* 24, 219–242.

Marston, S., Jones, J. and Woodward, K. (2005), 'Human Geography without Scale', *Transactions of the Institute of British Geographers* 30, 416–432.

Massey, D. (1999), 'Spaces of Politics', in Massey, D., Allen, J. and Sarre, P. (eds), *Human Geography Today* (Cambridge: Polity Press), 279–294.

Massey, D. (2001), 'Geography on the Agenda', *Progress in Human Geography* 25, 5–17.

Mauch, C., Stoltzfus, N. and Weiner, D. (eds) (2006), *Shades of Green: Environmental Activism around the Globe* (Lanham: Rowman and Littlefield).

McAfee, K. (1999), 'Selling Nature to Save It?' *Environment and Planning D: Society and Space* 7, 155–74.

McAfee, K. (2003), 'Neoliberalism on the Molecular Scale. Economic and Genetic Reductionism in Biotechnology Battles', *Geoforum* 34, 203–219.

McCarthy, J. (2005), 'Devolution in the Woods: Community Forestry as Hybrid Neoliberalism', *Environment and Planning A* 37: 995–1014.

McCarthy, J. (2006), 'Neoliberalism and the Politics of Alternatives: Community Forestry in British Columbia and the United States', *Annals of the Association of American Geographers* 96: 1, 84–104.

McCarthy, J. and Prudham, S. (2004), 'Neoliberal Nature and the Nature of Neoliberalism', *Geoforum* 35, 275–283.

McSweeney, K. (2004), 'The Dugout Canoe Trade in Central America's Mosqitia: Approaching Rural Livelihoods through Systems of Exchange', *Annals of the Association of American Geographers* 93: 3, 638–661.

Merchant, C. (1980), *The Death of Nature: Women, Ecology, and the Scientific Revolution* (San Francisco: Harper and Row).

Merchant, C. (1989), *Ecological Revolutions: Nature, Gender, and Science in New England* (Chapel Hill: University of North Carolina Press).

Mitchell, T. (2002), *Rule of Experts: Egypt, Techno-Politics, Modernit* (Berkeley: Uinversity of California Press).

Munch, R. (2007), *Globalization and Contestation: The New Great Counter-Movement* (London: Routledge).

Najam, A. (2005), 'Developing Countries and Global Environmental Governance: From Contestation to Participation to Engagement', *International Environmental Agreements: Politics, Law, and Economics* 5: 3, 303–321.

Neumann, R. (1998), *Imposing Wilderness: Struggles over Livelihood and Nature Preservation in Africa* (Berkeley: University of California Press).

Neumann, R. (2000), 'Land, Justice, and the Politics of Conservation of Tanzania', in Zerner, C. (ed.), *People, Plants, and Justice: The Politics of Nature Conservation* (New York: Columbia University Press), 117–133.

Neumann, R. (2005), *Making Political Ecology* (London: Hodder Arnold).

Nevins, J. (2003), 'Restitution over Coffee: Truth, Reconciliation, and Environmental Violence in East Timor', *Political Geography* 22, 677–701.

Peet, R. and Watts, M. (eds) (2004), *Liberation Ecologies: Environment, Development, Social Movements*, 2nd edition (London: Routledge).

Peluso, N. (1992), *Rich Forests, Poor People: Resource Control and Resistance in Java* (Berkeley: University of California Press).

Peluso, N. (1995), 'Whose Woods Are These?: Counter-Mapping Forest Territories in Kalimantan, Indonesia', *Antipode* 27: 4, 383–406.

Peluso, N. and Watts, M. (2001), *Violent Environments* (Ithaca: Cornell University Press).

Perreault, T. (2001), 'Developing Identities: Indigenous Mobilization, Rural Livelihoods, and Resource Access in Ecuadornian Amazonia', *Ecumene* 8: 4, 381–413.

Perreault, T. (2003a), 'Changing Places: Transnational Networks, Ethnic Politics, and Community Development in the Ecuadorian Amazon', *Political Geography* 22, 61–88.

Perreault, T. (2003b), '"A People with Our Own Identity": Toward a Cultural Politics of Development in Ecuadorian Amazonia', *Environment and Planning D: Society and Space* 21, 583–606.

Pickles, J. (ed.) (1995), *Ground Truth: The Social Implications of Geographic Information Systems* (New York: Guilford).

Reitan, R. (2007), *Global Activism* (London: Routledge).

Retort (2005), 'Blood for Oil?' *London Review of Books*. Available at: http://www.lrb.co.uk/v27/n08/reto01_.html.

Robbins, P. (2001), 'Tracking Invasive Land Cover in India, or Why Our Landscapes Have Never Been Modern', *Annals of the Association of American Geographers* 91: 4, 637–659.

Robbins, P. (2004), *Political Ecology: A Critical Introduction* (Oxford: Blackwell).

Robertson, M. (2000), 'No Net Loss: Wetlands Restoration and the Incomplete Capitalization of Nature', *Antipode* 32: 4, 463–493.

Robertson, M. (2004), 'The Neoliberalization of Ecosystem Services: Wetland Mitigation Banking and Problems in Environmental Governance', *Geoforum* 35, 361–373.

Robertson, M. (2006), 'The Nature That Capital Can See: Science, State, and Market in the Commodification of Ecosystem Services', *Environment and Planning D: Society and Space* 24, 367–387.

Rocheleau, D. and Edmunds, D. (1997), 'Women, Men and Trees: Gender, Power and Property in Forest and Agrarian Landscapes', *World Development* 25: 8, 1351–1371.

Rocheleau, D., Wangari, E. and Thomas-Slayter, B. (1996), *Feminist Political Ecology: Global Issues and Local Experiences* (London: Routledge).

Routledge, P. (2003), 'Convergence Space: Process Geographies of Grassroots Globalization Networks', *Transactions of the Institute of British Geographers* 28, 333–349.

Routledge, P. (2005), 'Reflections on the G8 Protests: An Interview with General Unrest of the Clandestine Insurgent Rebel Clown Army (Circa)', *ACME: An International E-Journal for Critical Geographies* 3: 2, 112–120.

Rutherford, S. (2007), 'Green Governmentality: Insights and Opportunities in the Study of Nature's Rule', *Progress in Human Geography* 31: 3, 291–307.

Schuurman, N. (2006), 'Formalization Matters: Critical GIS and Ontology Research', *Annals of the Association of American Geographers* 96: 4, 726–739.

Sheppard, E. (2005), 'Knowledge Production through Critical GIS: Genealogy and Prospects', *Cartographica* 40: 4, 5–21.

Smith, J. (2002), 'Globalizing Resistance: The Battle of Seattle and the Future of Social Movements', in Smith, J. and Johnston, H. (eds), *Globalization and Resistance: Transnational Dimensions of Social Movements* (Lanham: Rowman and Littlefield), 207–228.

Spronk, S. and Webber, J. (2007), 'Struggles against Accumulation by Dispossession in Bolivia: The Political Economy of Natural Resource Contention', *Latin American Perspectives* 34: 2, 31–47.

Star, L. (1991), 'Power, Technology, and the Phenomenology of Conventions: On Being Allergic to Onions', in Law, J. (ed.), *A Sociology of Monsters* (Oxford: Blackwell), 25–56.

Swyngedouw, E. (1997), 'Neither Global nor Local: "Glocalization" and the Politics of Scale', in Cox, K. (ed.), *The Spaces of Globalization: Reasserting the Power of the Local* (New York: Guilford), 137–166.

Swyngedouw, E. (2000), 'Authoritarian Governance, Power and the Politics of Rescaling', *Environment and Planning D: Society and Space* 18, 63–76.

The Independent (2006), 'Congo's Tragedy: The War the World Forgot', 5 May, available at: http://news.independent.co.uk/world/africa/article362215.ece.

Trouillot, M. (1995), *Silencing the Past: Power and the Production of History* (Boston: Beacon Press).

Watts, M. (2000), 'Political Ecology', in Sheppard, E. and Barnes, T. (eds), *A Companion to Economic Geography* (London: Blackwell), 257–274.

Watts, M. and McCarthy, J. (1997), 'Nature as Artifice, Nature as Artefact: Development, Environment, and Modernity in the Late Twentieth Century', in Lee, R. and Wills, J. (eds), *Geographies of Economies* (London: Arnold), 71–86.

Watts, M. and Peet, R. (2004), 'Liberating Political Ecology', in Peet, R. and Watts, M. (eds), *Liberation Ecologies: Environment, Development, Social Movements* (London: Routledge), 3–47.

Weiner, D. (1988), *Models of Nature: Ecology, Conservation, and Cultural Revolution in Soviet Russia* (Bloomington: Indiana University Press).

Weiner, D. (1999), *A Little Corner of Freedom: Russian Nature Protection from Stalin to Gorbachev* (Berkeley: University of California Press).

West, P. (2006), *Conservation Is Our Government Now: The Politics of Ecology in Papua New Guinea* (Durham: Duke University Press).

Whatmore, S. (2002), *Hybrid Geographies: Natures, Cultures, Spaces* (London: Sage).

Williams, M. (1994), 'The Relations of Environmental History and Historical Geography', *Journal of Historical Materialism* 20: 1, 3–21.

Wolford, W. (2005), 'Agrarian Moral Economies and Neo-Liberalism in Brazil: Competing World-Views and the State in the Struggle for Land', *Environment and Planning A* 37: 2, 241–261.

Worster, D. (1979), *Dust Bowl: The Southern Plains in the 1930s* (Oxford: Oxford University Press).

Worster, D. (1985), *Rivers of Empire: Water, Aridity, and the Growth of the American West* (New York: Pantheon Books).

Wright, A. and Wolford, W. (2003), *To Inherit the Earth: The Landless Movement and the Struggle for a New Brazil* (Oakland: Food First Publications).

Zerner, C. (ed.) (2000), *People, Plants, and Justice: The Politics of Nature Conservation* (New York: Columbia University Press).

PART 1
Translating Contentious Environmental Knowledge and Science

Chapter 2

The Contentious World of Jared Diamond's *Collapse*

Tim Forsyth

One of the most controversial developments in social science in recent years has been the analysis of the implicit politics contained within scientific statements (Jasanoff et al. 1995). For some critics, this trend has represented 'a betrayal of science and reason' (Ehrlich and Ehrlich 1996), or an irresponsible abandonment of logic in favour of postmodern excess (Levitt 1999). Yet, many social analysts of scientific knowledge do not wish to abandon a realist approach to understanding the world, but instead seek to show how the language and authority of 'science' have been used to defend and extend political values.

This chapter applies this kind of analysis to Jared Diamond's bestselling book about environmental change, *Collapse: How Societies Choose to Fail or Survive* (Diamond 2005). Diamond is a UCLA-professor who won a Pulitzer-Prize for an earlier book, *Guns, Germs and Steel* (Diamond 1997). Building on this success, *Collapse* started appearing in bookstores and airport terminals in 2005, and was lauded by many reviewers and journalists as essential environmental reading. According to the book's cover, the *New York Times* considered it 'magnificent', and two British Sunday newspapers claimed it to be 'riveting, superb, terrifying' (*Observer*), and 'a book that has to be read' (*Independent on Sunday*).

The basic argument of *Collapse* is that societies in both past and present have collapsed because of environmental degradation. Modern societies should therefore learn lessons from the past in order to avoid repeating such collapses. Many environmentalists have used this book to drive home messages about the need for society to live within ecological limits, and to acknowledge the dilemmas of current globalization, population growth and commercialization, especially in developing countries.

But how can we evaluate this book? This chapter applies an analysis of the scientific statements included within *Collapse* to argue that many of the book's apocalyptic claims are not rooted in scientific research, and are influenced instead by one clear political viewpoint. This analysis should not be taken to silence this viewpoint, nor to suggest there are no environmental problems. Rather, this chapter argues that the debate about environmental change is restricted by scientific assertions that reduce political debate. Instead, there is a need to be more aware of how political values underlie the gathering and representation of scientific knowledge.

What Does *Collapse* Say?

Collapse is about the history of human interactions with the natural world. The book's aim is to look empirically at the experiences of various historic (and sometimes quite recent) examples of human societies that have either failed catastrophically, or succeeded, because of their willingness to observe natural limits. These natural limits are usually taken to be the ecosystem services provided by soil, vegetation, and especially trees. The kinds of services provided include protection of the earth against excessive erosion; food and supplies from forests; and the overall regulatory functions provided by diverse vegetation on the hydrological cycle and nutrition of soil.

The book coins the word 'ecocide' to refer to the failure of societies to respect these ecosystem functions or natural limits. The usual means of ecocide lie in allowing human populations to grow so high as to transgress critical thresholds of land cover, or in exploiting soil and forest resources excessively as to prevent the overall maintenance of ecosystems. The book's overall lesson is to instruct current societies to respect these natural limits and consequently avoid similar catastrophe in the future.

Diamond presents evidence from various societies across the world. The book starts with a discussion of modern-day Montana, where problems resulting from the introduction of non-indigenous species and economic exploitation from mining are described. Next, the problems of past societies are listed from sites including Pacific islands (Easter, Pitcairn and Henderson); indigenous American settlements (Chaco, in the modern-US state of New Mexico, and Maya in today's Mexico); the Viking and Norse settlements of the northern Atlantic and Greenland); and New Guinea and Japan. Then, modern societies are discussed including the genocide in Rwanda; and comparisons of Haiti and the Dominican Republic; China and Australia.

Common themes are then identified. Diamond argues that historic societies faced eight categories of environmental threats: deforestation and habitat destruction; soil problems (including erosion, salinization, loss of nutrients); water management problems (frequently arising, he argues, from deforestation); overhunting; overfishing; introduced species (which upset previous ecosystems); human population growth to unsustainable levels; and increased impact per capita (or rates of consumption). And to these he adds four further threats faced by modern societies: anthropogenic climate change; accumulation of toxic chemicals; energy shortages; and future limits to the earth's photosynthetic capacity (a factor underlying accelerated climate change).

The most common themes, however, are the threats arising from deforestation and unchecked population growth. Diamond writes:

> like deforestation, soil problems contributed to the collapses of all past societies discussed in this book … (2005: 490).

And these problems are the root cause of various elements of political instability, which also impact on global imbalances between North and South:

problems of deforestation, water shortage, and soil degradation in the Third World foster wars there and drive legal asylum seekers and illegal emigrants to the First World from the Third World (498).

Two examples of historic and modern societies may illustrate these themes. Diamond focuses on the case of Easter Island, as evidence of deforestation-induced environmental collapse. According to evidence from travellers' records, and historical ecological pollen counts, Easter Island lost all trees above the height of 10 feet, and 90 per cent of its human population between the years 900 and 1722 (107). Diamond argues these changes are the result of foolish rates of deforestation on a site that was susceptible to erosion because of the erodibility of volcanic soils. He wrote:

> In short, the reason for Easter's unusually severe degree of deforestation isn't that those seemingly nice people were unusually bad or improvident. Instead, they had the misfortune to be living in one of the most fragile environments, at the highest risk for deforestation, of any Pacific people (118).

Yet, despite the alleged specificity of this example, Diamond then writes:

> The parallels between Easter Island and the whole modern world are chillingly obvious. Thanks to globalization, international trade, jet planes, and the Internet, all countries on Earth share resources and affect each other, just as did Easter's dozen clans (119).

And concerning genocide in Rwanda in 1994, Diamond argues that overpopulation and environmental degradation were important driving forces that are overlooked by most political analyzes of that crisis. He wrote:

> In recent decades, Rwanda and neighbouring Burundi have become synonymous in our minds with two things: high population, and genocide (313).

Diamond carefully avoids making the simplistic assertion that the genocide was purely the result of overpopulation or environmental degradation. But he argues that environmental pressures contributed to the genocide by listing evidence to suggest that land was scarce or that degradation was occurring.

> Friends of mine who visited Rwanda in 1984 sensed an ecological disaster in the making. The whole country looked like a garden and banana plantation. Steep hills were being farmed right up to their crests. Even the most elementary measures that could have minimized soil erosion ... were not being practiced (320).

Moreover, Diamond uses quotations from a French writer Gérard Prunier (1995) to make the statements that he clearly is uneasy saying himself, that the reason to kill was from a 'feeling there were too many people on too little land' (326, 328). Diamond then writes:

> I conclude that population pressure was *one* of the important factors behind the Rwanda genocide, that Malthus's worst-case scenario may sometimes be realized, and that Rwanda may be a distressing model of that scenario in operation. Severe problems of

overpopulation, environmental impact, and climate change cannot persist indefinitely: sooner or later they are likely to resolve themselves (328).

In his conclusion, Diamond uses the metaphor of a polder to indicate human life on earth: we are living on artificial means that require careful construction and maintenance. Despite this, Diamond calls himself a 'cautious optimist' (521) because he hopes humanity can learn to live within the natural limits of the earth, and that we can learn from historical mistakes.

What's Wrong with *Collapse*?

Diamond is a skilled writer, and his case studies provide fascinating reading. But his analytical argument is weakened by three questionable assumptions. The first is the presentation of natural limits (rather than social or economic factors) as the deciding influence on how societies collapse. The second is Diamond assumes a series of scientific causes-and-effects to explain human impacts that are more contested than he claims. Third, *Collapse* adopts a form of historical analysis that justifies modern moral decisions without acknowledging the social influences on how history is analyzed. We look at each in turn.

(i) Natural limits to growth

Collapse is clearly highly influenced by Malthusian thinking and the concept of overpopulation. This focus is attractive to many readers of the book. One reviewer (a US coral-reef ecologist) wrote:

> Since I am decidedly "coral-o-centric," I see the humans as the bad guys, and my standard response to the inevitable question of "What can we do to save the coral reefs?" has been "Tell people to move to Montana (and stop having babies)!" ... I encourage the author and his publisher to produce an abridged version that could be assigned to all college freshmen (Szmant 2006: 45–6).

Yet, what is surprising about this focus on overpopulation, and its glowing tributes from reviewers such as Szmant, is how these comments are made without reference to the long-standing debate about the significance of population as a driving force in environmental problems. Two points can be made here. First, *Collapse* tends to compress debates about overpopulation into a discussion of when it has happened and if it will happen again, rather than the *circumstances* in which overpopulation comes into existence, and how far processes of adaptation may reduce the significance of short-term scarcity. Diamond is rather selective on this point. He discusses the potential for technological optimism to overcome environmental problems (506), and points out the difficulties of making long-term technological transitions involving topics such as renewable energy. Yet, he does not discuss short-term re-allocations of technology or development assistance that can make adaptation more effective in developing countries. Research into natural hazards, for example, has indicated how countries with a low Human Development Index may encounter higher levels of mortality from events such as storms and drought, and that addressing social vulnerability may significantly reduce the hazardous nature of environmental events (Wisner et al. 2004). For example, research on vulnerability suggests that diversity of livelihood options, or political means of access of poor people to resources can mitigate absolute shortages of food or land (Sen 1981). For Diamond, however, there is no reference to these social dynamics of vulnerability (or how development interventions may reduce vulnerability) because for him the main sources of social vulnerability are natural limits.

Similarly, Diamond tends to avoid discussing different sides to debates about population that have been discussed for some years before his own research, and which are still hotly debated. For example, the 1994 United Nations Conference on Population in Cairo discussed the need for population control to be rooted in holistic approaches to development, including maternal health and women's reproductive rights. It also represented a radical change from discussions at the first United Nations Conference on Population at Bucharest in 1974 when the USA proposed the best solution to population growth was population control alone, in part influenced by Malthusian publications such as Paul Ehrlich's (1968) *The Population Bomb*. It is not surprising that, in his acknowledgements, Diamond offers a 'special badge of heroism' to Paul Ehrlich as one of six friends who critiqued his manuscript (526). (It is also likely that it was Paul and Ann Ehrlich whom Diamond cites as visiting Rwanda in 1984: see Ehrlich and Ehrlich 1996: 5–6. Ehrlich and Ehrlich (1996) have also dismissed social studies of environmental science.)

Second, Diamond gives overpopulation an active role in enhancing social collapse in excessive ways. This is most obviously seen in *Collapse*'s discussion of the genocide in Rwanda. Again, he is selective in his account. He uses the statements of Prunier (1995) to link environmental scarcity directly with genocide, and stresses that overpopulation is one cause. But despite such careful phrasing, the point of this discussion is to argue that environmental degradation and overpopulation led to genocide, and he refers to the Rwanda genocide as 'ecocide' in his introduction (7). Many observers would consider this explanation highly contentious because it overlooks the specific social contexts and history that contributed to disaster, rather than uniform environmental limits. Other authors have asked at what level does 'overpopulation' lead to crisis? Why are locations with similar population densities or environmental stress not experiencing genocide? How far does genocide actually reflect a lack of humanitarian intervention rather than some inevitable environmental cause? (Peluso and Watts 2001; Dallaire 2005). Genocide needs more complicated explanations than these, and Diamond is not helping by using Rwanda as a further chapter to support his Malthusian viewpoint.

(ii) Contested cause and effect

Throughout *Collapse*, Diamond uses compressed statements of environmental cause and effect to indicate how human actions have caused societies to collapse. In turn, he argues that avoiding (or retracting) these actions should prevent collapse. In particular, he considers deforestation to be the most common cause of environmental collapse. As one reviewer said in wholesale agreement (the same reviewer cited above):

> What kept coming to mind as I digested the various case studies was: deforestation, deforestation, deforestation! The destruction of forest cover appears to be the first and most important impact these extinct human societies inflicted on their environments: deforestation to clear land for farming, for building materials, for fuel (Szmant 2006: 44).

Again, this chapter does not argue deforestation has no impacts or should be ignored. But Diamond's descriptions of deforestation's impacts are startlingly simplistic and inaccurate. Various points can be made.

First, Diamond never defines deforestation other than the implicit assumption it involves cutting down trees. Yet, deforestation may vary widely from total clearfelling to selective cutting, to periodic patchwork clearance, to the careful alteration of tree species. All of these have different ecological impacts and social contexts, and consequently to refer to deforestation so generally is neither accurate nor helpful. Indeed, this point has been acknowledged for some decades among foresters and land-use analysts. In the words of one classic text:

> The generic term "deforestation" is used so ambiguously that it is virtually meaningless as a description of land-use change … It is our contention that the use of the term "deforestation" must be discontinued, if scientists, forest land managers, government planners and environmentalists are to have meaningful dialogue on the various human activities that affect forests and the biophysical consequences of those actions (Hamilton and Pearce 1988: 75).

Second, it is also clear that not just (in the words of Szmant) 'extinct societies' have conducted deforestation. Rather, much of North America and northern Europe were deforested centuries ago and have not collapsed. Clearly, deforestation is not as automatically degrading as either Diamond or Szmant suggest, and that context and impacts vary. This is true of locations where deforestation (and associated impacts) exist, rather than simply where alternative livelihoods may exist alongside land allegedly damaged by deforestation and erosion such as in Diamond's own example of Iceland where he claims deforestation caused erosion and precarious agriculture, but abundant fisheries allowed societies to avoid collapse.

Third, many of the links between deforestation and impacts on soil and water are now widely challenged by hydrologists or land-use analysts (Calder 1999; Bruijnzeel 2004; Ives 2004). Sometimes, these relationships are challenged because of the way erosion or deforestation are measured. For example, statistics concerning erosion under forest or crop cover vary according to whether sheet or gully erosion is measured. Sheet erosion – or the movement of soil across soil surfaces – is more likely to be higher under crops than under forests, where the forest canopy and floor litter protect soil. But gully erosion frequently occurs under forests as well as on cropland. The common model for measuring soil erosion, the so-called Universal Soil Loss Equation, is best used for sheet erosion, but has been criticized for overlooking the influence of local land-management, differential rainfall, or gully erosion in the tropics (Hallsworth 1987).

Fourth, much discussion of deforestation or soil erosion frequently includes assumptions about human actions that have been shown to be questionable. For example, research in West Africa has proposed that estimates of deforestation are exaggerated because they assume agriculture to be the only cause of deforestation, and fail to acknowledge local conservation practices or the role of dynamic natural fluctuations in forest cover (Fairhead and Leach 1996). Similarly, the assumption that population growth must cause farmers to cultivate steeper slopes has been questioned by research that shows how farmers appreciate the influence of slope on land failure (Forsyth 1996; Ives 2004). In short, this kind of research has indicated that popular explanations of environmental change tend to overstate and oversimplify human impacts, and simultaneously understate pre-existing natural causes and uncertainty (Thompson et al. 1986).

And fifthly, many of the proposed benefits of large-scale reforestation as a response to deforestation are also challenged. Reforestation (or afforestation) may vary as much as deforestation, but research has suggested that monoculture forest plantations may have negative impacts such as reducing biodiversity; enhancing erosion (by increasing splash erosion or reducing forest litter); and enhancing water shortages (by increasing evapotranspiration). Moreover, forest plantations remove land from potential agricultural uses, and hence may enhance local vulnerability to environmental or economic change.

As with simple statements about overpopulation, Diamond uses broad assumptions about the impacts of deforestation to indicate unquestioned linkages between human actions and environmental change. And again, it would be more constructive to discuss where, and under what circumstances, these impacts were problematic, rather than making universal assumptions.

(III) Coproduced values and history

A third characteristic is the lack of discussion in how historical or scientific 'facts' are used to justify arguments or moral judgments. *Collapse* is an overtly historical book. But the general purpose of its historical analysis is to justify the use of uncontested cause-and-effect statements, which are applied universally across time and space, rather than to analyze how varied rates of environmental change have become problematic in different contexts. In other words, Diamond offers no filter on how historical events are represented, but instead seems to see history as a hunting ground for evidence of scientific relationships he sees as unquestioned.

For example, Diamond says, 'The past offers us a rich database from which we can learn, in order that we may keep on succeeding' (3). But he demonstrates the teleology of this analysis by then writing, 'collapses for ecological or other reasons often masquerade as military defeats' (13). Concerning Easter Island, Diamond engages in historical analysis for *how* and *with whose knowledge* deforestation occurred as the key to understanding collapse. In historical terms, this is like searching a crime scene for the DNA of a chief suspect, rather than assessing what other evidence may exist.

The problem with this kind of selective historicism is that it further reifies the assumed cause-and-effect statements and the rigidity of "natural" limits discussed above. Moreover, presenting these causal relationships and limits as unquestioned also allows Diamond to justify a form of morality in his analysis of human behaviour that is also open to debate. This is most obvious in relation to his views about world order. For example, in *Collapse*'s conclusion, Diamond describes the spatial distribution of current political problems with areas of environmental stress. The two lists, he argues are identical, and comprise Afghanistan, Bangladesh, Burundi, Haiti, Indonesia, Iraq, Madagascar, Mongolia, Nepal, Pakistan, the Philippines, Rwanda, the Solomon Islands and Somalia, plus others (516). He writes:

> People in the Third World can now, intentionally or unintentionally, send us their own bad things: their diseases like AIDS, SARS, cholera, and West Nile fever ... unstoppable numbers of legal and illegal immigrants ... terrorists; and other consequences of their Third World problems (518). ... as a result of those problems of their own, [they] are also creating problems for us rich First World counties, which may end up having to provide foreign aid for them ... or may decide to provide them with military assistance to deal with rebellions and terrorists, or even have to send in our own troops (516).

Here, Diamond's worldview becomes most explicit. But are AIDS, immigration and terrorism specifically 'Third World' problems? Are they unnecessary burdens on richer countries? Are they caused by environmental stress?

Such views are obviously controversial. But linking these opinions to crude relationships between land-cover change and environment, or environmental limits and population, seems unacceptably simplistic for either political or environmental analysis. Yet, these statements are made in black and white in a book that many claim is both inspirational and accurate. Indeed, according to another review, an Australian consultant commented:

> *Collapse* ... is invigorating because ... you think, "I would have loved to have written this book!" – especially if you are a sustainability practitioner. ... Like a true scientist, [Diamond] postulates his hypothesis early and then sets out to prove it through supporting evidence (Jeyaretnam 2006: 42, 43).

But, clearly, *Collapse* is not as supported by scientific evidence as some would claim. And, its treatment of history seems designed to reify controversial beliefs about environment and world order. But why is *Collapse* considered important? And how can we improve environmental debate?

Making Sense of *Collapse*

It is tempting to approach *Collapse* by undertaking a more thorough survey of environmental evidence about the impacts of land-use-cover change. This approach would clearly be time consuming, and would include many aspects of analysis that are still ongoing and uncertain. For example, the diverse impacts of forest clearance could be discussed with more complexity, with greater nuance for possible environmental policies (e.g. see Bonell and Bruijnzeel 2004).

But what is worrying about this approach is that contrasting empirical evidence, and healthy environmental debate has already existed on the topics of human impacts for some years, but *Collapse* and its supporters have not reported it. Instead, perhaps a better approach might be to analyze the relationship of social values and environmental knowledge, which may explain how this book can be published when it makes statements that can be described as simplistic or outdated. This is the approach offered by social studies of scientific knowledge.

Political ecologists use the term environmental narratives to describe simplified explanations of environmental cause-and-effect that emerge in contexts where environmental knowledge and visions of social order are mutually dependent. The term 'narrative' has been used in literary analysis to refer to a storyline that gives meaning and structure to a sequence of events, often arranging them in terms of beginning, middle and end, sometimes with predefined functions for specific types of actor. Environmental narratives are similar because they give structure and purpose to diverse environmental events. But the interpretations given to narratives are always driven by a particular worldview or limited consultation of people (Roe 1991; Hajer 1995). Accordingly, narratives 'stabilize' complex and uncertain processes of environmental change into relatively simple and transferable summaries that also prescribe apparent solutions to environmental problems.

The contentious world of Jared Diamond – where 'Third World' populations transgress natural limits by triggering universal and unquestioned processes of environmental causality and political destabilization – can be described as an environmental narrative.

Two main approaches to environmental narratives have been proposed. First, Cultural Theory[1] (Thompson et al. 1990) argues that narratives reflect four key rationalities or worldviews in society. Worldviews result from two key criteria: the willingness to obey rules ('grid'), and the desire to act communally or individually ('group'). The four worldviews are consequently egalitarians (who believe in acting communally against rules); individualists (who dislike rules and groups); hierarchists (who want both rules and groups); and fatalists (who obey rules, but cannot act in groups). In turn, these worldviews are linked with environmental beliefs of environmental fragility; environmental resilience; the possibility of managing change; and a fatalistic attitude to change. (A further step is to map these respective positions onto well-known social actors such as NGOs, businesses, states and powerless people such as hill farmers).

Clearly, under this scenario, Jared Diamond may be called an egalitarian because he emphasizes global fragility and the need to control human growth within ecological limits. There are also elements of hierachism, because he urges 'cautious optimism' about developing rules to govern human behaviour. Alternative writers – who stress the benefits of economic growth, or the unnecessary restrictions of environmental policy – may be called individualists. Similar thinkers – such as those who write glowing book reviews despite the books' demonstrable shortcomings – also share appropriate worldviews, and consequently see alternative evidence as statistical aberrations.

---

1   Cultural Theory is based on the work of Mary Douglas, and uses a capital C and T to differentiate it from other theories.

The second approach to narratives has adopted a more poststructural emphasis by considering the historic and linguistic influences on how narratives are considered true. Hajer (1995), for example, calls narratives 'storylines' that have evolved over time as the result of how dominant groups or public debates have framed environmental change. Consequently, words such as 'deforestation' do not imply a universally defined process of biophysical change, but carry implicit notions of cause, effect, and social responsibility that reflect who helped shape them in the past. 'Deforestation' may also be called a 'hybrid' mixture of physical experience and human values, in the words of Bruno Latour (1993), which means that we are transferring historic and embedded social values when we use this term universally. Some of the poststructural criticism of narratives has also highlighted the problems of using positivist science as a basis for explaining environmental change, because it seeks to establish universal principles rather than identify the contexts under which different environmental changes become hazardous or meaningful. On this basis, Diamond is guilty of using simplified cause-and-effect statements to describe simplified processes, such as deforestation, in ways that do not acknowledge the diversity of these processes or their embedded social meanings that may not transfer between contexts.

But does this mean that deforestation has no impacts? Or that it is impossible to explain historic collapses in societies through environmental evidence? From a narrative perspective, the answers to these questions are still unsettled because there are various levels of uncertainty. It is uncertain how far environmental degradation caused impacts on historic societies. It is also uncertain how far we can use the experience of the past as a guide to the future.

Clearly, deforestation, erosion or other processes can cause environmental degradation. But the lesson of non-positivist approaches to environmental change is that the impacts of these changes vary according to the vulnerability and adaptive capacity of different societies. (This is how deforestation in Easter Island may have had more serious consequences than deforestation in, say, Minnesota.) Diamond is therefore probably right to discuss environmental degradation as a cause of collapse in some societies. But he is certainly wrong to claim that evidence of previous collapses is proof that erosion, deforestation or 'overpopulation' always lead to collapse, if there are no alternative livelihoods.

Moreover, Diamond claims that the lessons of the past indicate that we should organize environmental policy around reducing population growth, being aware of natural limits, and protecting forests. Yet, clearly, these opinions reflect his own worldview and use of historical narratives without consideration of how political intervention, development assistance, and reduction of vulnerability may achieve the same results without having to use contested notions of environmental change.

But does this mean Diamond is wrong to suggest there are global limits? Clearly, there may be limits to some populations that are vulnerable and lack adaptive capacity. But narrative analysis looks instead at how people claim there are limits. What political projects are supported by claiming natural limits? Second, which social groups define the limits? In *Collapse*, Diamond is dismissive about how aid or development may reduce vulnerability, and instead discusses natural limits and global threats. From a narrative perspective, all we can say is that Diamond likes to present environments as fragile and needing control for the same reasons he blames genocide on overpopulation, or sees terrorism, immigration and HIV/AIDS as 'Third World' problems imposed on richer countries (518). Does Diamond discuss natural limits, environmental fragility and overpopulation in order to legitimize a lack of development assistance?

Conclusion

This chapter has discussed Jared Diamond's *Collapse* in a critical manner in order to indicate how Diamond has made political arguments using scientific truth claims. Yet, rather than accepting these claims as factual and unchallenged, the chapter has argued that Diamond's own worldview has influenced how he has used scientific statements in order to justify a political position.

This argument, however, should not be taken as a suggestion that Diamond's political position should be dismissed, or not listened to. Moreover, it certainly should not be understood as arguing that societies should not be concerned about environmental change. Indeed, there are many reasons to be concerned. But we should be equally concerned about how we understand these potential problems, and how far political viewpoints shape how these relationships are discussed. Jared Diamond's *Collapse* argues in favour of political options today on the basis of scientific statements that need to be unpacked and seen to be less certain that he claims.

A crucial addendum to this argument is that we also need to identify room for political intervention in developing countries much more than Diamond suggests. Reading *Collapse* gives the impression that natural environmental limits are fixed, and that challenging these limits may result in disaster. An alternative viewpoint is that intervening to address social vulnerability – through practices such as long-term development, or short-term humanitarianism – may allow methods of avoiding disaster. This is most obvious in Diamond's analysis of genocide in Rwanda. Diamond's arguments suggests that the root causes of genocide lay in environmental scarcity. This argument forecloses debate about alternative political means to avoid genocide, which may involve a more pro-active involvement by richer countries in the long-term avoidance of social and political problems.

Diamond's environmental concerns are linked closely to a political worldview that dismisses the potential for development as a way to overcome both environmental stress and social vulnerability. If we are aware of these shortcomings – and hold Diamond to account – then this is a positive step for environmental policy.

References

Bonell, M. and Bruijnzeel, L. (2004), *Forests, Water, and People in the Humid Tropics: Past, Present, and Future Hydrological Research for Integrated Land and Water Management* (New York: Cambridge University Press).

Bruijnzeel, L. (2004), 'Hydrological Functions of Tropical Forests: Not Seeing the Soil for the Trees?' *Agriculture, Ecosystems and Environment* 104: 1, 185–228.

Calder, I. (1999), *The Blue Revolution: Land Use and Integrated Resource Management* (London: Earthscan).

Dallaire, R. (2005), *Shake Hands with the Devil: The Failure of Humanity in Rwanda* (New York: Arrow).

Diamond, J. (1997), *Guns, Germs and Steel: The Fate of Human Societies* (London: Jonathan Cape).

Diamond, J. (2005), *Collapse: How Societies Choose to Fail or Survive* (London: Penguin).

Ehrlich, P. (1968), *The Population Bomb* (New York: Ballantine Books).

Ehrlich, P. and Ehrlich, A. (1996), *Betrayal of Science and Reason: How Anti-Environmental Rhetoric Threatens our Future* (Washington, DC and Covelo, CA: Island Books).

Hajer, M. (1995), *The Politics of Environmental Discourse* (Oxford: Clarendon).

Hallsworth, E.G. (1987), *Anatomy, Physiology, and Psychology of Erosion* (New York: Wiley).

Hamilton, L, and Pearce, A. (1988), 'Soil and Water Impacts of Deforestation', in J. Ives and D. Pitt (eds), *Deforestation: Social Dynamics in Watershed and Mountain Ecosystems* (London: Routledge).

Fairhead, J. and Leach, M. (1996), *Misreading the African Landscape: Society and Ecology in a Forest-Savanna Mosaic* (Cambridge: Cambridge University Press).

Forsyth (1996), 'Science, Myth and Knowledge: Testing Himalayan Environmental Degradation in Northern Thailand', *Geoforum* 27: 3, 375–392.

Ives, J. (2004), *Himalayan Perceptions: Environmental Change and the Well-being of Mountain Peoples* (London: Routledge).

Jasanoff, S., Markle, G., Petersen, J. and Pinch, T. (eds) (1995) *Handbook of Science and Technology Studies* (Thousand Oaks: Sage).

Jeyaretnam, T. (2006), 'Book Review: Jared Diamond: Collapse', *Sustainability, Science and Policy* 2: 1, 42–3.

Latour, B. (1993), *We Have Never Been Modern* (Hemel Hempstead: Harvester Wheatsheaf).

Levitt, N. (1999), *Prometheus Bedeviled: Science and the Contradictions of Contemporary Culture* (New Brunswick: Rutgers University Press).

Peluso, N. and Watts, M. (eds) (2001), *Violent Environments* (Ithaca: Cornell University Press).

Prunier, G. (1995), *The Rwanda Crisis: History of a Genocide* (New York: Columbia University Press).

Roe, E. (1991), '"Development Narratives" or Making the Best of Blueprint Development', *World Development* 19: 4, 287–300.

Sen, A. (1981), *Poverty and Famines* (Oxford: Clarendon Press).

Szmant, A. (2006), 'Book Review: Jared Diamond: Collapse', *Sustainability, Science and Policy* 2: 1, 44–6.

Thompson, M., Warburton, M. and Hatley, T. (1986), *Uncertainty on a Himalayan Scale* (London: Milton Ash).

Thompson, M., Ellis, R. and Wildavsky, A. (1990), *Cultural Theory* (Boulder: Westview Press).

Wisner, B., Blaikie, P., Cannon, T. and Davis, I. (2004), *At Risk: Natural Hazards, People's Vulnerability and Disasters* (London: Routledge).

Chapter 3

Fight Semantic Drift!? Mass Media Coverage of Anthropogenic Climate Change

Maxwell T. Boykoff

Introduction

In November 2004, William Ruckelshaus – first United States Environmental Protection Agency (EPA) administrator – said, 'If the public isn't adequately informed [about climate change], it's difficult for them to make demands on government, even when it's in their own interest' (Ruckelshaus 2004). Previous research has found that the public in the United States (US) learns most of what it knows about science and policy from the mass media (Nelkin 1987, Wilson 1995).[1] More specifically, research has found that much of this understanding is gained from television and newspapers (Dunwoody and Peters 1992, Pew 2003). In discussing US mass media influence, Bennett states, 'Few things are as much a part of our lives as the news … it has become a sort of instant historical record of the pace, progress, problems, and hopes of society' (Bennett 2002: 10).

Meanwhile, over the past two decades, the United Nations-sponsored Intergovernmental Panel on Climate Change (IPCC) has brought together approximately 3,000 of the planet's top climate scientists, and it has enhanced understanding of global climate change through interpretation of emerging climate research via peer-reviewed and consensus-driven processes (Argrawala 1998b). With increasing confidence, the IPCC has asserted that climate change is a serious problem that has human (or 'anthropogenic') influences. Particularly, the Second Assessment Report from the IPCC in 1995 stated, 'the balance of evidence suggests that there is a discernible human influence on the global climate' (Houghton et al. 1995: 4). The Third Assessment Report that followed in 2001 went further by stating, 'There is new and stronger evidence that most of the warming observed over the last 50 years is attributable to human activities' (Houghton et al. 2001: 10). This detection of human influences was explained even more clearly in the 2007 IPCC Fourth Assessment Report (IPCC 2007). These 'critical discourse moments'

---

1 'Mass media' are considered the publishers, editors, journalists and others who constitute the communications industry and profession, and who produce, interpret and disseminate information, mainly through newspapers, magazines, television, radio and the Internet.

(cf. Carvalho 2005) have served to help solidify an emergent storyline of consensus in climate change.[2] D. James Baker, administrator of the US National Oceanic and Atmospheric Administration, has said about anthropogenic climate change that, 'There's a better scientific consensus on this than on any issue I know – except maybe Newton's second law of dynamics. ...' (Warrick 1997: A1).

While other research has focused analyses on how certain institutionalized discourses *within* climate science have shaped science-policy interactions (e.g. Pielke Jr. 2006; Boehmer-Christiansen 1994a, 1994b), this chapter surveys how this scientific discourse on anthropogenic climate change relates to mass media reporting in the US Here I briefly outline socio-political as well as bio-physical factors that challenge accurate media reporting on anthropogenic climate change, with a focus on the US – top emitter of greenhouse gases (GHGs), producing approximately 25 per cent of GHGs worldwide with just 4 per cent of the world's population. Through analyses of power (shaping knowledge, discourse and epistemic framings) and scale (both time and space), this work explores challenges that are faced in the mass media's portrayal of anthropogenic climate change. Through these inquiries, this chapter facilitates further work that reaches beyond modernist ontologies to examine discursive constructions that under gird the contemporary landscape.[3] Specifically, this work delineates how US media representational practices have depicted conflict rather than coherence regarding scientific explanations of anthropogenic climate change over time. In other words, this chapter outlines how the convergent views within science regarding anthropogenic climate change have been depicted as contentious when reported through US mass media, thus giving an appearance of increased uncertainty and 'debate'. I briefly argue that these challenges have increased contentions at the interface of science and policy, and have been detrimental to ongoing discourse on policy action regarding climate change.[4] These explorations aim to provide better traction for paths towards problem resolution.

---

2   In other related issues, such as the *rate* of temperature change, no clear consensus has yet emerged. In the piece 'Middle Stance Emerges in Debate over Climate', Revkin provides a nuanced view of a number of these differences as well as characterizing a 'middle-ground' populated by 'non skeptical heretics' while acknowledging consensus on anthropogenic climate change (2007).

3   This draws from analogous terms of 'heterogeneous constructionism' from Demeritt (2001) and co-production (see Murdoch 1997a, 1997b; Jasanoff 2004). The first term focused on the politics surrounding this heterogeneous construction of 'global warming', recognizing that meaning is constructed and manifested through *both* the ontological conditions of nature, and contingent social and political processes involved in interpretations of this nature.

4   Boehmer-Christiansen writes, 'The environmental policy problem facing societies is twofold' (1) how to ensure that useful knowledge informs policy without being misused and distorted; and (2) how to respond to this knowledge in the context of existing patters of interests, perceptions and commitments (1994).

International Environmental Politics and Policy Negotiations

Climate change is widely considered one of the most pressing crucial environmental, social and political problems now facing the planet.[5] In March of 2003, United Nations Weapons Inspector Hans Blix said, 'To me the question of the environment is more ominous than that of peace and war ... I'm more worried about global warming than I am of any major military conflict' (Norris 2003). Moreover, in January of 2005, Rajendra Pachauri – chair of the IPCC – said that the world has 'already reached the level of dangerous concentrations of carbon dioxide in the atmosphere' and called for immediate and 'very deep' cuts in greenhouses gas emissions if humanity is to survive (Lean 2005).

Meanwhile, research has revealed that media coverage of climate change has played a significant role in translations between science, policy and the public (e.g. Smith 2005; McComas and Shanahan 1999; Bell 1994). Moreover, an appearance of a storyline of increased uncertainty and 'debate' has arisen in public and policy discourse through such troubles in translating the consensus in climate science regarding anthropogenic influences. Despite the aforementioned scientific consensus regarding human contributions to climate change, the US Bush administration has been ambivalent in regards to this scientific view. In the summer of 2006, CNN broadcast a statement by US President George W. Bush that, 'I have said consistently that global warming is a problem. There is a debate over whether it is man-made or naturally caused' (2006).[6] However, in August 2004 the Bush administration released a report indicating that rising temperatures in North America were attributable in part to human activity and that this was having detectable effects on animal and plant life. US media coverage of the report noted that this was 'a striking shift in the way that Bush administration has portrayed the science of climate change' (Revkin 2004a: A18). Amid seemingly contradictory rhetoric, the Bush administration has not yet engaged in global emissions reductions programs. In this high-stakes arena of US federal administration action on climate change, science and policy have engaged in dialectical interactions – both complementary and contentious – while US mass media report on them.[7]

---

5   The US mass media uses the terms 'climate change' and 'global warming' interchangeably. It is also used this way extensively in policy and public discourse. However, it should be noted that strict scientific definitions make a distinction between these two terms: 'Global warming' refers to a more specific facet of climate change, which is the climate characteristic of temperature. Even more specifically, this refers to increases in temperature over time. 'Climate change' also accounts for changes in other climate characteristics, such as rainfall, ice extent and sea levels.

6   This can also be juxtaposed with the statement from US President George H.W. Bush, who said in 1990, 'We all know that human activities are changing the atmosphere in unexpected and in unprecedented ways' (Compton 1990).

7   For example, Bush Administration interventions in the climate change section of the EPA's draft 'Report on the Environment' in 2003 demonstrated highly politicized contestations taking place on the terrain of climate policy making. Ultimately, the EPA deleted the entire section on climate change, because 'as agencies wrote in an internal April 29th memo, the

The Pursuit (and Translation) of 'Truth'?

Considerations of human activities' role in changing the climate have generated much attention. While it can be challenging to appropriately characterize and delineate general views in a broadly construed 'scientific community', the particular collaboration of top scientists from around the world on climate change presents a unique opportunity to do so (see Argrawala 1998a). Edwards and Schneider have stated that the IPCC is 'generally regarded as the single most authoritative source of information on climate change and its potential impacts on environment and society' (1997: 1–2). Adger et al. have explored different climate change discursive regimes, and have described a 'managerial discourse' as one that draws primary authority from scientific findings, focuses on macro-scale solutions, and bases actions on external policy interventions (2001). This work thus concentrates on the IPCC as a group that effectively articulates a 'global environmental management' discourse (e.g. Houghton et al. 1990; Watson et al. 1997; McCarthy et al. 2001), one that interacts with national and international policy discourse.[8] Such solidified discourse on anthropogenic climate change has tethered institutional activities and actors[9] to 'storylines' (Hajer 1993, 1995), and has reproduced itself (or has sought to do so) through policy-relevant research statements and decisions over time.

Taylor and Buttel posited that the organizational arrangements that define what are environmental problems (such as anthropogenic climate change) can be seen 'simultaneously (as) a scientific framework and a movement ideology' and is thus 'particularly vulnerable to deconstruction' (1992: 406). Others have pointed out that as scientific understanding improves, rather than settling questions, it often unearths new and more questions to be answered. Moreover, *greater* scientific understanding actually can contribute to *more* complicated policy decision-making by offering up a greater supply of knowledge from which to develop and argue varying interpretations of that science (e.g. Rayner 2006, Sarewitz 2004). In other words, anytime that the biophysical is captured and categorized at the science-practice interface, it undergoes varying degrees of politicized interpretation, as influenced by power and scale via temporal and spatial contexts (Jasanoff and Wynne 1998). Thus, in the discourse assembled by the IPCC, a certain way of viewing things is privileged, and gains salience. In the case of anthropogenic climate change, the stakes within and between carbon-based industry and society are high. In combination, the science-practice interface has become a contentious discursive battlefield, and one particularly important for intervention through geography approaches (Burgess 2005).

A number of scholars through time have looked at different aspects of the politicized landscape within which the scientific authority (such as the IPCC) has

---

White House's watered-down language "no longer accurately represents scientific consensus on climate change"' (Gregory 2003).

8   It also incorporates assessments from salient climate research that informs IPCC statements (e.g. Santer et al. 1996; Tett et al. 1999; Allen et al. 2000; Easterling et al. 2000; Crowley 2000; Karl and Trenberth 2003) on anthropogenic climate change.

9   For instance, the United Nations Framework Convention on Climate Change Conference of Parties Meetings (UNFCCC COPs), or James Hansen, climate scientist for NASA Goddard Institute for Space Studies (GISS).

arisen and is maintained as well as negotiated at the interface with policy and practice. Frameworks that have been developed adopt a range of perspectives, from a straightforward 'deficit model' of science-policy communications where scientific findings fill a 'knowledge deficit' for policy use (see Gross 1994) to a more complex 'parliament of things' where facts-values interact with nature-society (see Latour 2004). The spectrum of views offers a spectrum of advantages as well as drawbacks. For instance, in the case of the 'deficit model', it has been critiqued for being too simple a characterization of the dynamic interactions therein, as well as overly adopting a vision of science as open, universal and objective (e.g. Oreskes 2004). The latter Latourian perspective has been comprised of sometimes impenetrable and obfuscated considerations, thus limiting its potential for application (e.g. Castree 2006). While this chapter does not delve centrally into these varied perspectives, it is important to acknowledge such subjective complexities when interrogating US coverage of scientific explanations of anthropogenic climate change. In a 2006 editorial 'what drives environmental policy?' Rayner grapples with these epistemological challenges and states:

> For good or ill, we live in an era when science is culturally privileged as the ultimate source of authority in relation to decision-making. The notion that science can compel public policy leads to an emphasis on the differences of viewpoint and interpretation within the scientific community. From one point of view, public exposure to scientific disagreement is a good thing. We know that science is not capable of delivering the kinds of final authority that is often ascribed to it. Opening up to the public the conditional, and even disputatious nature of scientific inquiry, in principle, may be a way of counteracting society's currently excessive reliance on technical assessment and the displacement of explicit values-based arguments from public life. However, when this occurs without the benefit of a *clear understanding* of the importance of the substantial areas where scientists *do* agree, the effect can undermine public confidence (2006: 6, emphasis added).

The focus here is on how 'clear understanding' in science – scientific consensus on anthropogenic climate change – has been framed by US mass media rather as contention and conflict, thus affecting public as well as policy confidence.[10] This makes it therefore possible to reconcile the unavoidably politicized science illustrated by that of IPCC with this distinct facet of climate change. In other words, this consensus is not the 'truth' translated, but signifies an aspect of climate change where there is clear understanding. Demeritt has noted this ongoing tension at the interface of climate science-practice. He states that, 'the notion of a purely scientific realm of objective facts as distinct from a political one of contestable values is idealized by nearly all participants in debates about climate change, even as it is habitually breached in ordinary practice' (2006: 472). Therefore, in 'ordinary' practice, policy-relevant work of the IPCC embodies multiple views of the role of science in policy.

---

10 Framing is an inherent part of cognition, defined as the ways in which elements of discourse are assembled that then privilege certain interpretations and understandings over others (Goffman 1974). The process of media framing involves an inevitable series of choices to cover certain events within a larger current of dynamic activities. These events are then converted into news stories.

On one hand, there exists an element of deficit model processes, as the assembly of top climate science in IPCC reports does fill a knowledge gap that proves useful for policy considerations. On the other hand, the 'facts' that emerge in these documents are also influenced by values and perspectives at the human-environment interface. Therefore, while viewing IPCC deliberations as imperfect and IPCC reports as heterogeneous constructs of facts and values, this analysis remains useful in order to track concatenate movements of convergence (in climate science) and confusion (through media coverage of anthropogenic climate change).[11]

Contested Spaces and Anthropogenic Climate Change at the Media-Science-Policy Interface

Translations of meaning through multiple scales and dimensions of power

Professional and disciplinary practices – such as news production – make actors both the object of discipline and the instruments of its exercise (Foucault 1984). Like discourse, Foucault reasons that power is only partly derived from the artifact – here as the news story – constructed by history (Foucault 1976). Conversely, news stories are partly generated from power relationships developed and through a rich history of professionalized journalism (Starr 2004). Socio-political and economic factors that buttress journalism has given rise to distinct norms and values (Bennett 2002). This mobilization of power is subtle and consequent ruptures are complex and contradictory. Specifically, discontinuities arise in climate change media coverage through professional journalistic norms and values often put in place to safeguard against potential abuses of asymmetrical power. In fact, discontinuities can arise in media coverage of anthropogenic climate change through the very professional journalistic norms and values that have developed to safeguard against potential abuses of asymmetrical power.

Epistemic framings – and therefore construction of meaning – are then derived through combined structural and agential components. Asymmetrical influences, also feedback into these social relationships and further shape emergent frames of 'news', knowledge and discourse. While some factors like access through ownership and control are more readily apparent, other influences, such as journalists' training and nature's agency are more concealed. The power that arises from these elements then becomes re-embedded in *macro*-relations, such as decision-making in a capitalist political economy, and *micro*-processes such as everyday journalistic practices. These complex and iterative processes contribute to what is then considered 'systemic power' (Stone 1980), and affect how anthropogenic climate change coverage is produced and disseminated. As Foucault writes, 'power produces knowledge' and 'it is not the activity of the subject of knowledge that produces a corpus of knowledge,

---

11 This has been referred to elsewhere as 'risks of communication'. In an analysis of climate change discourse in science, politics and mass media in the German context, the authors write, 'the assumption that scientific knowledge is communicated unequivocally to the rest of society and is transformed into actions according to a one-dimensional rationality is no longer tenable (if it ever was)' (Weingart et al. 2000: 280).

useful or resistant to power, but power-knowledge, the processes and struggles that transverse it and of which it is made up, that determines the forms of possible domains of knowledge' (1979: 27–28). In other words, these dynamic interactions of multiple scales and dimensions of power critically contribute to how anthropogenic climate change coverage is produced and disseminated.

Figure 3.1 shows the distribution of the newspaper articles and television news segments by year, from 1988 through 2006, and provides a useful look into how the issue gained salience over time. Particular increases in coverage took place in 1990, 1992, 1997, 2001-2002, and 2006 (see Boykoff and Boykoff (2007) for a detailed discussion regarding fluctuations in coverage over time).

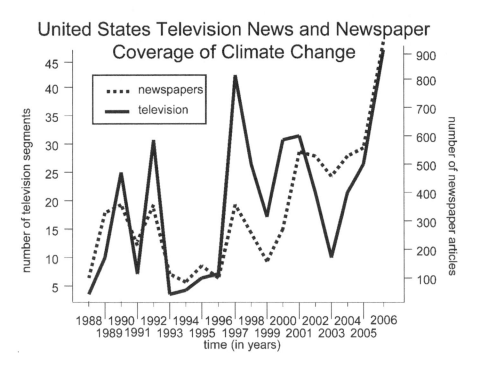

**Figure 3.1   US newspaper articles and television segments on
global warming/climate change, 1988–2006**

Note: The newspapers analyzed here are the *New York Times*, *Los Angeles Times*, *Washington Post* and *Wall Street Journal*. The television programs analyzed are ABC World News Tonight, CBS Evening News and NBC Nightly News.

At the *macro* political-economic level, in recent years, media organizations have continued to consolidate power. Efficiency and profit increasingly influence news production (Bennett 1996), and a number of studies have explored how economic pressures and ownership structures have affected news content (Herman and

Chomsky 1988, Parenti 1986). McChesney has asserted, 'The core structural factors that influence the nature of media content include the overall pursuit of profit, the size of the firm, the amount of direct and indirect competition facing the firm and the nature of that competition, the degree of horizontal and vertical integration, the influence of advertising, the specific interests of media owners and managers, and, to a lesser extent, media employees. In combination these factors can go a long way to providing a context (and a trajectory) for understanding the nature of media content …' (1999: 31). In other research, Gilens and Hertzman have provided 'systematic evidence that financial interests of media owners influence not only newspaper editorials but straight news reporting as well' (2000: 383). Additionally, deadlines and space considerations constrain journalists (Schudson 1978), as do editorial preferences and pressure from publishers (Schoenfeld et al. 1979). The amount of exposure and placement (front page or buried deep in the newspaper), as well as the use of headlines and photographs – which are often editorial decisions – can also affect how events and situations are construed by the public (Parenti 1997). Also, quick deadlines can lead to one-source stories (Dunwoody 1986). Economic considerations have led to decreased mass-media budgets for investigative journalism (McChesney 1999), and has led to more journalists working as 'generalists' by covering many areas of news, rather than 'specialists' on a particular news beat (Gans 1979, Bennett 2002). According to research by Dunwoody and Peters, the typical journalist in the US is 'even less likely to have majored in science or math than is the average US resident' (1992: 208). This lack of science training among journalists can lead to mistranslations of anthropogenic climate change.

Moreover, these different communities have developed varying conceptions time-scale in their professionalized cultures, and this affects communications. In climate science, new insights are gained through longer-term iterative endeavours such as field research, modelling, and peer-review processes. In climate policy, political cycles, negotiations and mobilization of constituencies generally function in short- to medium-time scales. In journalism, 'breaking news', efficiency and profitability often pressure journalists to work on short-term time scales. Also in terms of time-scale, structural constraints also play a critical role in hampering more effective communications between communities via the media. For example, in climate science – and more broadly, academia – most reward systems are structured such that little is gained professionally through increased 'non-academic' pursuits such as media outreach. Geophysicist Henry Pollack has written, '[T]here is an underlying feeling that in an academic career, advancement is retarded by spending time in nonacademic endeavours' (2003: 75). Conversely, much can be risked in terms of being misquoted about the implications of one's research. In media and policy communities, there is immediacy in the need for policy-relevant information, and these disparate priorities lead to communication breakdowns. Also, that corrections in media reporting– crucial to the precision of climate science – are placed in following days without much prominence is disconcerting for ongoing interactions between communities. Overall, these time-scale discrepancies contribute to continued troubles in translation of anthropogenic climate change.

These examples begin to work across scales from macro-level political economic factors to *micro*-level processes such as journalistic norms intersect with these elements and shape news content (Jasanoff 1996). Journalistic norms include objectivity, fairness, accuracy and balanced reporting. The quotidian micro-practices of journalism can serve to amplify asymmetrical power through providing coverage to this minority viewpoint. Studies have found that through the journalistic norm of balance, US mass media have provided informationally biased coverage of anthropogenic climate change (Boykoff 2008, Boykoff and Boykoff 2004). For example, a *Los Angeles Times* article entitled 'Studies Point to Human Role in Warming' contributed to this notion of a lively 'debate', and contained the following passage:

> The issue of climate change has been a topic of intense scientific and political debate for the past decade. Today, there is agreement that the Earth's air and oceans are warming, but disagreement over whether that warming is the result of natural cycles, such as those that regulate the planet's periodic ice ages, *or* caused by industrial pollutants from automobiles and smokestacks (McFarling 2001: A1, emphasis added).

As another example, a *Wall Street Journal* article focused on conflict without context, effectively deemphasizing and confusing scientific consensus on anthropogenic climate change. In an article entitled, 'Weighing the Evidence of Global Warming: A Scientist's Work on Ocean Becomes Fodder for Skeptics – Much to his Dismay', the author writes:

> For much of his 30-year career, oceanographer Lloyd Keigwin yearned to publish something groundbreaking, a paper that would win him the recognition of his fellow marine geologists. His wish came true in 1996 with his pioneering research on ocean temperatures and climate, which gave scientists a glimpse of natural warming and cooling patterns in the recent past. But his data made an even bigger splash with global-warming skeptics – including oil giant Exxon Mobil Corp – who cited his work on Web sites and other venues as evidence that natural climate change, rather than human activity, is causing the Earth to warm (Herrick 2001: B1).

In television news, an *NBC Nightly News* report began a piece on climate change by covering a statement by US president George Bush that warming might be due to anthropogenic causes. However, the news story then counterbalanced Bush's statement with quotes from commentator Rush Limbaugh and explicit mention of a potential 'upside: growing seasons in northern areas could last longer, crops and timber become more plentiful and therefore cheaper, fuel bills could be lower, and there could be more warm weather for construction or recreation' (Hager 2002). Despite that the story began through coverage of Bush's policy position aligning momentarily with the international scientific community, the story in effect engendered conflict through these counterweights.

In terms of environmental issues more broadly, Freudenberg discusses this conception of power enabling privileged constructions of '*non*-problematicity'. Through embedded power and leveraged legitimacy, 'if one person or social group is able to obtain privileged access to valued resources without having other persons or groups challenge that privilege – or perhaps even notice it – so much the better' (Freudenberg 2000: 106). In the case of anthropogenic climate change, oppositions have arisen primarily in a cohesive group of 'climate contrarians' – also called 'skeptics' – who have gained greater discursive traction through the media, as a result, affecting public understanding. Climate contrarians include scientists S. Fred Singer,[12] Robert Balling,[13] Sallie Baliunas,[14] Richard Lindzen[15] and Patrick Michaels,[16] and policy actors James Inhofe,[17] Chuck Hagel[18] and James Schlesinger,[19] among others (Beder 2002; Gelbspan 1998). This opposition speaks out stridently against

---

12   Fred Singer, former professor at the University of Virginia, has not published original research material on climate change for over two decades. He is executive director of the Science and Environmental Policy Project (a think-tank established in 1990), and has served on the advisory board for the conservative group the Advancement of Sound Science Coalition. He has been a consultant for Exxon, Shell, Unocal, Sun Oil and Arco.

13   Robert Balling, from Arizona State University, is on a panel of scientific advisors for the Greening Earth Society, and has been on the advisory council for the Information Council on the Environment and has worked with the Global Climate Coalition and the Competitive Enterprise Institute. He has worked for the Information Council on the Environment (ICE) – a creation of a group of utility and oil companies. Balling has also received funding for his research work from groups such as the German Coal Mining Association and Cyprus Minerals.

14   Dr. Sallie Baliunas is on the panel of scientific advisors for the Greening Earth Society. She has chaired the scientific advisory board of the George C. Marshall Institute, has provided research for the Competitive Enterprise Institute and the Cooler Heads Coalition, and has gone on a media tour questioning the science of ozone depletion and global warming paid for by the climate-skeptical group the Global Climate Coalition.

15   Dr. Richard Lindzen is a professor of meteorology at the Massachusetts Institute of Technology. He has served on the advisory board for the conservative George C. Marshall Institute, and has been a consultant to the fossil fuel industry (Beder 2002; Gelbspan 1998).

16   Dr. Patrick Michaels, from the University of Virginia, edits the Western Fuels Association-funded *World Climate Report*. He is on advisory boards for the conservative groups TASSC, and the Greening Earth Society. In 2005 he wrote a book entitled 'Meltdown: The Predictable Distortion of Global Warming by Scientists, Politicians and the Media', published by Cato Institute Books. Michaels has received funding for his research from groups such as the Edison Electric Institute, The Intermountain Rural Electric Association, and the Western Fuels Association (Beder 2002; Gelbspan 1998).

17   Until recently, Inhofe (Republican – Oklahoma) had been the chair of the Senate Environment and Public Works committee. He has accepted over $1 million in contributions from natural resources and energy interests since he was elected to US Senate in 1994. On the Senate floor, he said, 'could it be that man-made global warming is the greatest hoax ever perpetrated on the American people? It sure sounds like it' (Inhofe 2003).

18   In 1998, Chuck Hagel (Republican – Nebraska) held up a petition on the Senate Floor that was signed by 17,000 'scientists' and questioned the link between climate change and human activities. Upon closer analysis, many of the names were not scientists, and included Ginger Spice of the Spice Girls as well as the fictitious Benjamin 'Hawkeye' Pierce from M.A.S.H., and Perry Mason (in Johansen 2002: 92).

19   Schlesinger – former Director of the Central Intelligence Agency, former Secretary of Defense under Presidents Nixon and Ford and Secretary of Energy under President Carter – has been a member of the Board of Directors for Peabody Energy. He has written op-eds on climate change in the *Washington Post* entitled 'Climate Change: The Science Isn't Settled' (Schlesinger 2003) and *Los Angeles Times* entitled 'Cold Facts on

the aforementioned consensus in climate science, and through this privileged access and power, has amplified uncertainty on human contributions to climate change. Through media coverage, this has also consequently shaped public perception of a lively dispute where there is little. Research by McCright and Dunlap has focused on this opposition movement, and has examined how certain individuals work to develop competing discourses that disempower top climate science, and effectively gain a foothold in national and international discourse on the causes of climate change (2000, 2003; see also McCright 2007).

For example, in *New York Times* coverage of US president George W. Bush's 2001 withdrawal from Kyoto Protocol negotiations, climate contrarian Fred Singer was quoted:

> Dr. S. Fred Singer, a retired physicist and longtime critic of research indicating a warming trend with a human cause, said he hoped Mr. Bush would kill the treaty outright. "The Kyoto Protocol is like a vampire", Dr. Singer said. "You need to drive a stake through its heart. Otherwise it'll keep coming back and causing problems" (Revkin 2001: A7).

As another example, the *Washington Post* covered the emergent IPPC consensus statement on a 'discernible human influence', but offered a counterweight to these top climate scientists. The article quoted astrophysicist Piers Corbyn: 'As far as we are concerned, there's no evidence for global warming, and by the year 2000 the man-made greenhouse theory will probably be regarded as the biggest scientific gaffe of the century' (Atkinson 1995: A10).

Overall, these everyday practices of journalism take shape through socialization and feedbacks via asymmetrical power relationships. As discussed above, this is more usefully considered as a complex and dynamic relationship (e.g. Rayner 2006; Latour 2004; Collins and Evans 2002; Newell 2000), rather than a one-way communication of science-as-tool-to-inform-policy considered in positivist conceptions (e.g. Gross 1994; Roberts 2004). While science and policy clearly shape media reporting and public understanding, it is also the case that journalism and public concern shape ongoing climate science and policy decisions. The perspective offered here seeks to unpack the terms of the 'debate' to which the media contributed; where knowledge, meaning and discourse are seen as negotiated on a field of struggle (Hall 1988). This approach interrogates how power and scale construct, reflect and reveal heterogeneous and complex phenomena such as language, knowledge and discourse (Forsyth 2003).

While covering this politicized arena, media representations have in fact presented a contentious picture of present scientific understanding of anthropogenic climate change. This has taken place at both *macro* and *micro*-scales, where unequal power relations intersect with bio-physical processes. Elsewhere, Fitzsimmons and Goodman have appropriately critiqued research approaches that consider nature as merely the backdrop upon which heterogeneous human actors contest and battle for epistemological and material successes (1998). With anthropogenic climate change, assessments from the IPCC and elsewhere have interpreted biophysical

---

Global Warming: Officials Mustn't Be Unduly Influenced by Scare Tactics that Cite Climate Change' (Schlesinger 2004).

processes through a scientific focus on changes in the *mean* of particular climate characteristics over time. For example, estimations of future temperature changes on the planet are widely considered through average global atmospheric temperature readings. In the IPCC's 2001 report, climate scientists placed the expected global mean temperature increase in the range of 2.5°F (1.4°C) to 10.4°F (5.8°C) by 2100 (Houghton et al. 2001). This and other similar estimates of mean temperature change have been picked up by US mass media and have been included in numerous news reports on anthropogenic climate change over time. However, through a focus on changes in global *averages*, this epistemic framing runs an increased risk in climate policy decision-making by minimizing considerations of potential non-linear and abrupt climate changes (Mastrandrea and Schneider 2004; Schneider et al. 2002). Consequently, media coverage of nature's agency – as mediated through scientific research – in response to human influences is then often subsumed by socio-political and economic concerns, such as how certain GHG reduction efforts may constrict economic activities. With such socio-political concerns at the fore, greater stress is placed on the danger of climate policy on trade and the economy, rather than also considering how trade and the economy may have detrimental effects on the global climate. Such partial readings of the complex problem then inevitably and preemptively constrict policy considerations.

Furthermore, challenges in translation are exacerbated by the complex bio-physical nature of climate change itself, seen by *New York Times* Environment Reporter Andrew Revkin as 'the classic incremental story' (2004c). Media studies researchers have asserted that, 'Journalists are less adept at reporting complex phenomena ... (and) have difficulty reporting stories that never culminate in obvious events' (Fedler et al. 1997: 94). While scientific insights regarding complex issues such as anthropogenic climate change evolve over years and decades, through journalistic norms and pressures, media tends to take 'snapshot' selections from this steady stream of enhanced understanding. Through positivist epistemologies and concatenate framing, 'these approaches seek no analytical engagement with nature, and thus silently re-affirm its objectification in accordance with the ontology of orthodox social science' (Goodman 1999: 25). Therefore, proposed movements towards solutions are limited through this 'epistemic truncation', by not considering a range of both socio-political and biophysical factors that effectively widen possibilities for proposed actions and resolution.

Conclusion

Climate scientists have commented that we are now in the 'Anthropocene Era' (Falkowski et al. 2000). For billions of years, radiative forcing – producing changes in the climate – has occurred in response to energy imbalances. Atmospheric temperature change – considered climate change's 'fingerprint' (Wigley 1999) – is the most evident form of climate forcing. Indeed, separating and distinguishing anthropogenic forcing from natural variability is a challenging task, but over time, signal (anthropogenic) to noise (natural) schemes have been derived. Through such work, human activities have been found to account for the majority of global warming since 1850 (Tett et al. 1999, Allen et al. 2000, Crowley 2000). In addition to the aforementioned IPCC statements of 1995 and 2001, numerous findings and statements have also bolstered this position. For instance, a 2004 study published in *Science* found that of 928 articles on global climate change that were published in peer-reviewed scientific journals between 1993 and 2003, not a single paper opposed this consensus position (Oreskes 2004). The research also found that, 'all major scientific bodies in the United States whose members' expertise bears directly on the matter have issued similar statements' (Oreskes 2004: 1686).

However, a focus on natural variation has still been the crux of the current US federal administration's climate change discourse. President George W. Bush has focused future federally funded climate research on the natural components of climate change rather than anthropogenic forcing. This research agenda – covered extensively in the US mass media – has run counter to many international climate scientists' assertions that the focus of climate research should be on the human component of climate change (Revkin 2004b). Bush administration ties to carbon-based energy interests notwithstanding, this shift from a research focus to that of natural variation represents significant hegemonic efforts to configure ontological foundations of complex nature-society relations. Concomitantly, it challenges the epistemological bases of discourse and recommended action rooted in international scientific undertakings. As discussed above, the dualist separation of nature and society serve to further entrench modernist and positivist ontological discussions and prematurely foreclose possible alternatives. In short, such measures – furthered by coverage in the US mass media – have stymied discourse on other possibly meaningful mitigative and adaptive climate change actions.

This chapter begins to outline how and why US media representational practices have depicted conflict and contentions rather than coherence regarding scientific explanations of anthropogenic climate change. It briefly shows that challenges faced in US media coverage of anthropogenic climate change are not random. Rather, they are systemic and occur in two main and interrelated ways: first, through complex socio-political and economic reasons rooted in macro-power relations, as well as micro-processes that under gird professional journalism; and second, through innate biophysical characteristics that contradictorily shape knowledge, discourse and epistemic framings at multiple scales. So, rather than dismissing US climate change policy positions as impetuous or as mere manifestations of the power of vested interests, actions can instead be seen as products of the complex interweaving of competing threads of meaning. And, this meaning is maintained and contested through feedbacks of these intertwined socio-political and biophysical processes.

Lessons learned extend beyond the issue of anthropogenic climate change to other contentious geographies, recognizing that meaning is constructed and manifested through *both* the ontological conditions of nature, and contingent social and political processes involved in architectures that shape differing interpretations of this nature (Haraway 1997). This work nests itself into larger 'cultural circuits' of climate change reflection and action (Carvalho and Burgess 2005), that is itself nested in multi-scale socio-political and biophysical influences. While some researchers have suggested that possibilities for problem resolution would be enhanced through a re-focusing on politics and policy (e.g. Pielke 2004, Sarewitz 2004), such efforts must also account for influences by non-state actors such as mass media. When the process of framing – whereby meanings are constructed and reinforced – confuse rather than clarify scientific understanding of anthropogenic climate change via the media, this can create spaces for US federal policy actors to defray responsibility and delay action regarding climate change. In sum, this chapter is a step towards elucidating how the situated and influential role of the US mass media has generated public perception of lively and contentious debate amid convergent views in anthropogenic climate science.

References

Adger, W.N., Benjaminsen, T.A. et al. (2001), 'Advancing a Political Ecology of Global Environmental Discourses', *Development and Change* 32, 681–715.

Allen, M.R., Stott, P.A. et al. (2000), 'Quantifying the Uncertainty in Forecasts of Anthropogenic Climate Change', *Nature* 407, 617–620.

Argrawala, S. (1998a), 'Context and Early Origins of the Intergovernmental Panel on Climate Change', *Climatic Change* 39, 605–620.

Argrawala, S. (1998b), 'Structural and Process History of the Intergovernmental Panel on Climate Change', *Climatic Change* 39, 621–642.

Atkinson, R. (1995), 'Reaching a Consensus is the Hot Topic at Global Climate Conference', *Washington Post*, Washington, DC: A10, March 28.

Beder, S. (2002), *Global Spin: The Corporate Assault on Environmentalism* (White River Junction, VT: Chelsea Green Publishing Company).

Bell, A. (1994), 'Media (Mis)Communication on the Science of Climate Change', *Public Understanding of Science* 3, 259–275.

Bennett, W.L. (1996), 'An Introduction to Journalism Norms and Representations of Politics', *Political Communication* 13, 373–384.

Bennett, W.L. (2002), *News: The Politics of Illusion* (New York: Longman).

Boehmer-Christiansen, S. (1994a), 'Global Climate Protection Policy: The Limits of Scientific Advice, Part 1', *Global Environmental Change* 4: 2, 140–159.

Boehmer-Christiansen, S. (1994b), 'Global Climate Protection Policy: The Limits of Scientific Advice, Part 2', *Global Environmental Change* 43, 185–200

Boykoff, M.T. and Boykoff, J.M. (2004), 'Bias as Balance: Global Warming and the US Prestige Press', *Global Environmental Change* 14: 1, 125–136.

Boykoff, M.T. (2008), 'Lost in Translation? United States Television News Coverage of Anthropogenic Climate Change, 1995–2004', *Climatic Change* 86: 1, 1–11.

Boykoff M.T. and Boykoff, J.M. (2007), 'Climate Change and Journalistic Norms: A Case-study of US Mass-media Coverage', *Geoforum* 38: 6, 1190–1204.

Burgess, J. (2005), 'Follow the Argument Where it Leads: Some Personal Reflections on "Policy-relevant" Research', *Transactions of the Institute of British Geographers* 30, 273–281

Carvalho, A. (2005), 'Representing the Politics of the Greenhouse Effect: Discursive Strategies in the British Media', *Critical Discourse Studies* 2: 1, 1–29.

Carvalho, A. and Burgess, J. (2005), 'Cultural Circuits of Climate Change in UK Broadsheet Newspapers, 1985–2003', *Risk Analysis* 25: 6, 1457–1469.

Castree, N. (2006), 'A Congress of the World', *Science as Culture* 15, 159–170.

Chouliaraki, L. and N. Fairclough (1999), *Discourse in Late Modernity: Rethinking Critical Discourse Analysis.* (Edinburgh: Edinburgh University Press).

Crowley, T.J. (2000), 'Causes of Climate Change Over the Past 1000 Years', *Science* 289: 14 July, 270–277.

CNN (2006), 'Headline News: Live Today', 26 June.

Collins, H.M. and Evans, R. (2002), 'The Third Wave of Science Studies: Studies of Expertise and Experience', *Social Studies of Science* 32, 235–296.

Compton, A. (1990), 'Environment, Global Warming and Bush', *ABC World News Tonight*, February 5.

Demeritt, D. (2006), 'Science Studies, Climate Change and the Prospects for Constructivist Critique', *Economy and Society* 35: 3, 453–479.

Demeritt, D. (2001), 'The Construction of Global Warming and the Politics of Science', *Annals of the Association of American Geographers* 91: 2, 307–337.

Dunwoody, S. (1986), 'The Science Writing Inner Club: A Communication Link Between Science and the Lay Public', in S.M. Friedman, S. Dunwoody and C.L. Rogers, *Scientists and Journalists: Reporting Science as News* (New York and London: The Free Press), 155–169.

Dunwoody, S. and Peters, H.P. (1992), 'Mass Media Coverage of Technological and Environmental Risks', *Public Understanding of Science* 1: 2, 199–230.

Easterling, D.R., Meehl, G.A. et al. (2000), 'Climate Extremes: Observations, Modeling, and Impacts', *Science* 289: 5487, 2068–2074.

Edwards, P.N. and Schneider, S.H. (1997), 'The 1995 IPCC Report: Broad Consensus or "Scientific Cleansing"?', *Ecofable/Ecoscience* 1: 1, 3–9.

Falkowski, P., Scholes, R.J. et al. (2000), 'The Global Carbon Cycle: A Test of Our Knowledge of Earth as a System', *Science* 290: 5490, 291–296.

Fedler, F., Bender, J.R. et al. (1997), *Reporting for the Media* (Fort Worth, TX: Harcourt Brace College Publishers).

FitzSimmons, M. and Goodman, D. (1998), 'Incorporating Nature: Environmental Narratives and the Reproduction of Food', in B.B.N. Castree, *Remaking Reality: Nature at the Millennium* (Routledge), 194–220.

Forsyth, T. (2003), *Critical Political Ecology: The Politics of Environmental Science* (London: Routledge).

Foucault, M. (1976), *The Archaeology of Knowledge* (New York: Harper and Row).

Foucault, M. (1979), *Discipline and Punish: The Birth of the Prison* (New York: Harper and Row).

Foucault, M. (1984), 'Space, Knowledge and Power', in P. Rabinow, *The Foucault Reader* (New York: Pantheon Books).

Freudenburg, W.R. (2000), 'Social Construction and Social Constrictions: Toward Analyzing the Social Construction of "The Naturalized" and well as "The Natural"', in G. Spaargaren, Mol, A.P.J. and Buttel, F.H. *Environment and Global Modernity* (London: Sage), 103–119.

Gans, H. (1979), *Deciding What's News* (New York: Pantheon).

Gelbspan, R. (1998), *The Heat is On: The Climate Crisis, the Cover-up, the Prescription* (Boston: Perseus Press).

Gilens, M. and Hertzman, C. (2000), 'Corporate Ownership and News Bias: Newspaper Coverage of the 1996 Telecommunications Act', *Journal of Politics* 62: 2, 369–386.

Goffman, I. (1974), *Frame Analysis: An Essay on the Organization of Experience* (London: Harper and Row).

Goodman, D. (1999), 'Agro-food Studies in the "Age of Ecology": Nature, Corporeality, Bio-politics', *Sociologia Ruralis* 39: 1, 17–38.

Gregory, D. (2003), 'Environment: EPA Report', *NBC Nightly News*, 19 June.

Gross, A. (1994) 'The Roles of Rhetoric in the Public Understanding of Science', *Public Understanding of Science* 3, 3–23

Hager, R. (2002), 'In Depth – Bush, Global Warming Policy Change', *NBC Nightly News*, 3 June.

Hall, S. (1988), *The Hard Road to Renewal: Thatcherism and the Crisis of the Left* (London: Verso).

Hajer, M. (1995), *The Politics of Environmental Discourse: Ecological Modernization and the Policy Process* (Oxford: Clarendon Press).

Hajer, M. (1993), 'Discourse Coalitions and the Institutionalization of Practice: The Case of Acid Rain in Britain', Fischer, F. and Forester, J. (eds), *The Argumentative Turn in Policy Analysis and Planning* (Durham and London: Duke University Press).

Haraway, D. (1997), *Modest_Witness@Second_Millennium. FemaleMan© Meets OncoMouse™ Feminism and Technoscience* (New York and London: Routledge).

Herman, E.S. and Chomsky, N. (1988), *Manufacturing Consent: The Political Economy of the Mass Media* (Toronto, Canada: Pantheon Books).

Herrick, T. (2001), 'Weighing the Evidence of Global Warming: A Scientist's Work on Ocean Becomes Fodder for Skeptics – Much to his Dismay', *Wall Street Journal*, New York, NY: B1. March 22.

Houghton, J.J., Filho, L.G.M. et al. (eds) (1996), *Climate Change 1995: The Science of Climate Change* (Cambridge: Cambridge University Press).

Houghton, J.T., Ding, Y., Griggs, D.J., Noguer, M., van der Linden, P.J. and Xiaosu, D. (2001), *Climate Change 2001: The Scientific Basis* (Geneva, Switzerland: IPCC).

Houghton, J.T., Jenkins, G.J. et al. (1990), *Climate Change: The Scientific Assessment* (Cambridge: IPCC).

Inhofe, J. (2003), 'The Science of Climate Change Senate Floor Statement', *Chair, Committee on Environment and Public Works*, US Senate, 28 July.

IPCC (2007), 'Climate Change 2007: The Physical Science Basis, Summary for Policymakers' (Geneva, Switzerland).

Jasanoff, S. (1996), 'Science and Norms in Global Environmental Regimes', in Hampson, F.O. and Reppy, J. *Earthly Goods: Environmental Change and Social Justice* (Ithaca and London: Cornell University Press), 173–197.

Jasanoff, S. and Wynne, B. (1998), 'Science and Decisionmaking', Rayner, S. and Malone, E.L. (eds), *Human Choice and Climate Change: The Societal Framework* (Columbus, OH: Battelle Press), 1–87.

Jasanoff, S. (ed.) (2004), *States of Knowledge: The Co-production of Science and Social Order* (London: Routledge).

Johansen, B.E. (2002), *The Global Warming Desk Reference* (Westport, CN: Greenwood Press).

Karl, T.R. and Trenberth, K.E. (2003), 'Modern Global Climate Change', *Science* 302: 5 December, 1719–1723.

Latour, B. (1994), *Politics of Nature: How to Bring the Sciences into Democracy* (Cambridge, MA: Harvard University Press).

Lean, G. (2005), 'Global Warming Approaching Point of No Return, Warns Leading Climate Expert', *The Independent*, London, 23 January.

Mastrandrea, M.D. and Schneider, S.H. (2004), 'Probabilistic Integrated Assessment of "Dangerous" Climate Change', *Science* 304: 23 April, 571–575.

McCarthy, J.J., Canziani, O.F. et al. (2001), *Climate Change 2001: Impacts, Adaptation and Vulnerability* (Geneva, Switzerland: IPCC).

McChesney, R.W. (1999), *Rich Media, Poor Democracy: Communication Politics in Dubious Times* (Urbana and Chicago: University of Illinois Press).

McComas, K. and Shanahan, J. (1999), 'Telling Stories about Global Climate Change: Measuring the Impact of Narratives on Issue Cycles', *Communication Research* 26: 1, 30–57.

McCright, A.M. and Dunlap, R.E. (2000), 'Challenging Global Warming as a Social Problem: An Analysis of the Conservative Movement's Counter-Claims', *Social Problems* 47: 4, 499–522.

McCright, A.M. and Dunlap, R.E. (2003), 'Defeating Kyoto: The Conservative Movement's Impact on US Climate Change Policy', *Social Problems* 50: 3, 348–373.

McCright, A.M. (2007), 'Dealing with Climate Change Contrarians', Moser, S.C. and Dilling, L. (eds), *Creating a Climate for Change: Communicating Climate Change and Facilitating Social Change* (Cambridge: Cambridge University Press), 200–212.

McFarling, U.L. (2001), 'Studies Point to the Human Role in Global Warming', *Los Angeles Times*, Los Angeles, CA: A1, 13 April.

Murdoch, J. (1997a), 'Towards a Geography of Heterogeneous Associations', *Progress in Human Geography* 21: 3, 321–337.

Murdoch, J. (1997b), 'Inhuman/nonhuman/human: Actor-network Theory and the Prospects for a Nondualistic and Symmetrical Perspective on Nature and Society', *Environment and Planning D: Society and Space* 15, 731–756.

Nelkin, D. (1987), *Selling Science: How the Press Covers Science and Technology* (New York: W.H. Freeman).

Newell, P. (2000), *Climate for Change: Non-state Actors and the Global Politics of the Greenhouse* (Cambridge: Cambridge University Press).

Norris, J. (2003), 'Interview with Hans Blix. USA', *Music Television (MTV)*, 13 March.

Oreskes, N. (2004), 'Beyond the Ivory Tower: The Scientific Consensus on Climate Change', *Science* 306: 5702, 1686.

Oreskes, N. (2006), 'Science and Public Policy: What's Proof Got to Do with It?', *Environmental Science and Policy* 7, 369–383.

Parenti, M. (1986), *Inventing Reality: The Politics of the Mass Media* (New York: St. Martin's Press).

Parenti, M. (1997), 'Methods of Media Manipulation', C. Jensen (ed.), *Twenty Years of Censored News* (New York, Seven Stories Press).

Pew Research Center for People and the Press (2003), Summary Report, 21 October.

Pielke, R.A. (2004), 'When Scientists Politicize Science: Making Sense of Controversy over "The Skeptical Environmentalist"', *Environmental Science and Policy* 7, 405–417.

Pielke, Jr. R.A. (2006), 'Misdefining "Climate Change": Consequences for Science and Action', *Environmental Science and Policy* 8, 548–561.

Pollack, H. (2003), 'Can the Media Help Science?' *Skeptic* 10: 2, 73–80.

Rayner, S. (2006), 'What Drives Environmental Policy?' *Global Environmental Change* 16: 1, 4–6.

Revkin, A. (2001), 'Bush's Shift Could Doom Air Pact, Some Say', *New York Times*, New York: A7, 17 March.

Revkin, A. (2004a), 'US Report, In Shift, Turns Focus to Greenhouse Gases', *New York Times*, A18, 26 August.

Revkin, A. (2004b), 'Bush vs. Laureates: How Science Became a Partisan Issue', *New York Times*, New York, NY, 19 October.

Revkin, A. (2004c), 'From a Discussion in the "Science Reporting" Course at the Columbia University School of Journalism', New York, NY, 7 October.

Revkin, A. (2007), 'Middle Stance Emerges in Debate over Climate', *New York Times*, A16, 1 January.

Roberts, J. (2004), *Environmental Policy* (London: Routledge).

Ruckelshaus, W. (2004), *Journalists/Scientists Science Communications and the News Workshop* (organizers: Anthony Socci and Bud Ward), University of Washington, 8–10 November.

Santer, B.D., Taylor, K.E. et al. (1996), 'A Search for Human Influences on the Thermal Structure of the Atmosphere', *Nature* 382, 39–46.

Sarewitz, D. (2004), 'How Science Makes Environmental Controversies Worse', *Environmental Science and Policy* 7, 385–403.

Schlesinger, J. (2003), 'Climate Change: The Science Isn't Settled', *Washington Post*, Washington, DC, 7 July.

Schlesinger, J. (2004), 'Cold Facts on Global Warming: Officials Mustn't Be Unduly Influenced by Scare Tactics that Cite Climate Change', *Los Angeles Times*, Los Angeles, 17B, 22 January.

Schneider, S.H., Rosencranz, A. et al. (eds) (2002), *Climate Change Policy: A Survey* (Washington, DC: Island Press).

Schoenfeld, A.C., Meier, R.F. et al. (1979), 'Constructing a Social Problem: The Press and the Environment', *Social Problems* 27: 1, 38–61.

Schudson, M. (1978), *Discovering the News: A Social History of American Newspapers* (New York: Basic Books).

Smith, J. (2005), 'Dangerous News: Media Decision Making about Climate Change Risk', *Risk Analysis* 25: 6, 1471–1482.

Starr, P. (2004), *The Creation of the Media: Political Origins of Modern Communications* (New York: Basic Books).

Stone, C.N. (1980), 'Systemic Power in Community Decision Making: A Restatement of Stratification Theory', *American Political Science Review* 74, 978–990.

Taylor, P.T. and Buttel, F.H. (1992), 'How Do We Know We Have Global Environmental Problems? Science and the Globalization of Environmental Discourse', *Geoforum* 23: 3, 405–416.

Tett, S.F.B., Stott, P.A. et al. (1999), 'Causes of Twentieth-Century Temperature Change Near the Earth's Surface', *Nature* 399, 569–572.

Warrick, J. (1997), 'The Warming Planet; What Science Knows', *The Washington Post*, Washington, DC, A1, 12 November.

Watson, R.T., Zinyowera, M.C. et al. (1997), *The Regional Impacts of Climate Change: An Assessment of Vulnerability* (Cambridge: IPCC).

Weingart, P., Engels, A. and Pansegrau, P. (2000), 'Risks of Communication: Discourses on Climate Change in Science, Politics, and the Mass Media', *Public Understanding of Science* 9, 261–283.

Wigley, T.M.L. (1999), *The Science of Climate Change: Global and US Perspectives* (Washington, DC: Pew Center on Global Climate Change).

Wilson, K.M. (1995), 'Mass Media as Sources of Global Warming Knowledge', *Mass Communications Review* 22: 1 and 2, 75–89.

Chapter 4

Whose Scarcity? The *Hydrosocial* Cycle and the Changing Waterscape of La Ligua River Basin, Chile

Jessica Budds

Introduction

This chapter examines a conflict over water resources for irrigation in a rapidly developing agricultural valley in Chile's semi-arid *Norte Chico*. It explores escalating demand for new water resources, in particular groundwater, for export-oriented fruit plantations, and its implications in terms of water resources management and access to water rights between commercial and peasant farmers. Situated within the broadly-defined political ecology tradition, the chapter draws on emerging theorizations of human-nature relations to analyze how the nature of the conflict is shaped by social power, discourse and nature's agency, as well as how the conflict, and attempts to address it through the production of a physical hydrological assessment, configure uneven socio-ecological outcomes at the basin scale.

The chapter starts by outlining a political ecology approach to environmental change. The first section reviews recent theories of 'hybrid' or 'social' nature, that further attempts to conceptualize nature as simultaneously social and material, and proceeds to consider emerging critical perspectives on environmental science, that question both its supposed neutrality and its role in producing 'facts' to underpin policy. This section finishes by presenting recent applications of these perspectives to water, through the concept of the *hydrosocial* cycle, that simultaneously considers the physical hydrological cycle and the ways in which water is also controlled and shaped by social power relations and institutions, and which forms the analytical framework for the empirical case. The following section presents the case study of the material and discursive conflict over water resources in La Ligua river basin, focusing in particular on competing representations of water scarcity and visions of solutions to local water problems. The section then evaluates a hydrological assessment that was undertaken to respond to this situation, and the socioecological implications of the resulting water allocation. The penultimate section analyzes how social construction, discourse and scale are implicated in the conflict and its responses, and how they are reflected in the changing social relations and waterscape of the valley. The final section draws some conclusions about the dialectical relationship between the materiality of water and the social relations of control over it, and the very real implications for peasant farmers in La Ligua.

The Politicization of Nature and Environmental Change

Departing from the premise that socio-ecological change has *political* underpinnings, which occur at different spatial and temporal scales, this section draws on recent theorizations of nature-society relations, as well as perspectives that critique environmental science and place greater attention on the agency of biophysical processes, to explore the relationship between social power and control over water.

Political ecology departs by recognizing that conventional technical approaches to natural resources (engineering, economics, law, resource management, science) are inadequate for explaining the complexity of environmental change. Such approaches are limited by their consideration of the environment as an assemblage of physical components that are subject to human manipulation. This forms the basis of 'human-environment impact' analyses, which focus on how human actions modify the natural environment. These conventional approaches are problematic in two key ways. First, they give little consideration to the complexity and interrelatedness of the social dimensions of environmental change, and instead tend to identify immediate spatial and temporal causes, with less attention to wider and/or multiple factors. Second, their primary explanations are often based on simple cause-effect relationships between human activity and environmental change, which are frequently regarded as self-evident, rather than the result of careful assessment (see also Forsyth, this volume). Failing to look beyond the 'observable' boundaries of environmental problems results in a depoliticized and dehistoricized analysis that fails to fully capture the complex nature of society-environment dynamics, and typically orients remedial measures towards these 'symptoms' rather than their 'causes' (Bryant and Bailey 1997; Castree and Braun 2001; O'Riordan 1999; Paulson 2003).

Political ecology enquiry has responded by seeking to understand the 'complex metabolism between nature and society' (Johnston et al. 2000: 590). In particular, it has more closely examined the roles of different social groups and institutions in society-nature relations, their vested interests and the power relations between them, and how these shape often uneven social and ecological outcomes, across wider spatial and temporal scales (Blaikie 1985; Bryant and Bailey 1997; Castree and Braun 2001; Paulson and Gezon 2005; Robbins 2004; Zimmerer and Bassett 2003). Power relations, which are by definition unequal, play a role in determining *how* nature is transformed: *who* exploits resources, under *which* regimes and with *what* outcomes for both social fabrics and physical landscapes (Bryant and Bailey 1997; Swyngedouw 1997b).

Given the often competing interests among different social actors *vis-à-vis* environmental management, power relations must be exercised to be effective. This is achieved by 'socially constructing' nature, whereby nature is perceived in distinct ways by different actors, within particular moments and contexts, and consequently represented according to these positionalities. The various constructions are then mobilized through associated discourses, through which social actors frame issues (definitions, problems, solutions) and promote them in ways that coincide with their particular interests and visions of how nature should be managed (Blaikie 1995, 2001; Braun and Wainwright 2001; Castree 2001b; Demeritt 2001). Political ecologists have thus sought to question conventional understandings and deconstruct situated constructions of nature, in order to uncover the power structures underlying them (Castree 2001a, 2001b).

Scale has been an important aspect of political ecology research, principally in relation to considering political economic influences on environmental change beyond the local level and the present time (Blaikie 1985). However, conventional notions of scale – 'local', 'national' and 'global' – have been criticized as preconceived divisions of space within which social processes occur, and have given rise to fresher notions of scale as more horizontal, complex, diverse, dynamic and socially produced (Mansfield 2005; Marston et al. 2005; Swyngedouw 1997a). In addition, ecological scales, such as the watershed, have largely been neglected, thus raising the dual challenges of working beyond conventional divisions of space and integrating social and ecological scales (Zimmerer and Bassett 2003).

The view of environmental issues as politicized, constructed and discursive is simultaneously challenged and complemented, by two theoretical developments: hybrid or social nature and critical approaches to environmental science.

Social nature

The *a priori* separation of 'nature' and 'society' into two distinct domains – the foundation of environmental studies, sciences and management – has been identified as both artificial and problematic (Castree 2001b; Escobar 1999; Haraway 1991; Harvey 1996; Latour 1993). As a result, attempts have been made to reconceptualize nature and society as a 'hybrid' (Swyngedouw 2004; Whatmore 2002), 'social nature' (Blaikie 2001; Castree 2001a, 2001b) or 'socio-nature' (Swyngedouw 1997b).[1] This resonates with Harvey's (1996) dialectical approach, which transcends the materiality of nature by instead considering it to be constituted, and reconstituted, by the *processes* that continually transform it:

> Dialectical thinking emphasizes the understanding of processes, flows, fluxes and relations over the analysis of elements, things, structures, and organized systems ... [these] do not exist outside of or prior to the processes, flows and relations that create, sustain or undermine them (Harvey 1996: 49).

A dialectical understanding of nature emphasizes the two-directional dynamics of social and natural processes in socioecological change. This allows nature *itself* to be reconceptualised as inescapably politicized, rather than merely the object of political processes, thus overcoming the dualistic perspective of nature as external to social power. In this way, a hybrid perspective enables the political processes and power relations that underlie fused 'socioecological' change to be elucidated, as power and socioecological change can be understood as mutually and dialectically constitutive (Castree 2001b; Harvey 1996; Paulson et al. 2003). This rejects the view of nature as a purely material domain over which policies are made and social struggles occur, to an integrated 'social nature' in which the agency of non-human natures also shapes social power (Braun and Wainwright 2001; Castree 2001b; Whatmore 2002).

---

1   Neologisms used elsewhere include 'quasi-object' (Latour 1993), 'cyborg' (Haraway 1991; Swyngedouw 2004), and 'imbroglio' (Swyngedouw 2004; Whatmore 2002).

Rethinking environmental science

An acceptance that no presentation of 'reality' can be free from the positionality or discourse of the actor articulating it carries profound and far-reaching implications: it not only questions the very production of environmental knowledges and values, but also sheds scrutiny on the institutions that produce such 'truths' and 'facts'. While such ideas have critiqued the supposedly objective role of the state in resource allocation, these had already been addressed by Foucault's theory of governmentality, which holds that government technical approaches to environmental management tacitly coincide with the interests of powerful groups (politicians, technocrats, capitalists) (Foucault 2002). However, more recently, they have prompted scrutiny of the validity of science to provide knowledge about how nature works and how best to manage it (Castree 2001b; Demeritt 1998, 2001). This comprises two aspects: the validity of environmental science in producing facts, and the agency of biophysical processes in environmental change.

Moving beyond long-standing criticisms of positivism, work on the politics of environmental science has explored the constructions of nature that underlie science, and the values and interests of scientists (Forsyth 2003; Robbins 2004; Zimmerer and Bassett 2003). This has led to a reassessment of the superiority and neutrality of scientific knowledge in explaining ecological processes, and, in turn, has called into question the validity of science as a basis for environmental management (Braun and Wainwright 2001; Demeritt 1998; Forsyth 2003; Zimmerer and Bassett 2003). The use of science to produce 'facts' about nature, and as a basis for policymaking, is problematic. Many scientific assessments exclude social processes or make generalized assumptions about the human causes of environmental degradation, especially by failing to disaggregate the actions of different social groups. Environmental policies that accept such definitions and assessments can result in measures that not only fail to address the underlying causes, but also penalize groups who make minimal contributions to the problem. This calls for a more comprehensive analysis of scientific concepts and analysis into political ecology research, that seeks to uncover both the sociopolitical framings of problems and the epistemological limitations of 'facts' (Forsyth 2003; see also Forsyth, this volume).

Early political ecology work contained little explicit discussion of ecology, and tended to consider nature as both inert and the *object* of environmental struggles, which resulted in the predominance of largely political or political economy explanations for environmental change. Such analyses of environmental change or degradation are problematic because they may overlook both the complexities of ecological 'reality' and the agency of biophysical processes (Forsyth 2003; Walker 2005; Zimmerer and Bassett 2003). Notions of a more complex 'ecological' reality underlying environmental change are contentious. These stem from contemporary, but contested, debates within some ecological sciences that question notions of ecological equilibrium in favor of alternative nonlinear theories (such as chaos theory), which posit that environmental behavior is more complex, less uniform and more multi-scale than presented in conventional science (Forsyth 2003). The key debate centers on whether irregularities observed in natural processes are attributable to randomized behavior or the limitations of scientists' ability to observe and measure them, especially beyond the local scale.

The hydrosocial cycle

Drawing on these related traditions, water has been reconceptualised from a purely material 'resource', that is tangible and observable, and which can be quantified, harnessed and manipulated, to a socio-natural one:

> Water is a "hybrid" thing that captures and embodies processes that are simultaneously material, discursive and symbolic (Swyngedouw 2004: 28).

Hybrid water can thus be understood as a dynamic flow that circulates within the *hydrosocial* cycle, as opposed to the physical hydrological cycle. As well as examining how water flows through the physical environment (atmosphere, surface, subsurface, biomass), the hydrosocial cycle also considers how water is manipulated by social actors and institutions (culture, laws, modes of management, hydraulic works, industries) (Bakker 2003; Swyngedouw 2004). Transcending the materiality of water allows the social power relations that are embodied within hydrosocial change to be revealed. These will be apparent from the different ways in which water is used by diverse social actors and its different modes of management, which are always embedded within space and time. The water-power nexus, in turn, will configure its socioecological outcomes, which will be reflected in both the physical waterscape and the social relations of access to, and exclusion from, water (Swyngedouw 1997b, 1999, 2004). Given that water is a strategic and essential resource for most economic activities, and that more powerful actors will strive to control it, water management should therefore not be merely understood as the partitioning and allocation of resources among different users, but as an inherently political struggle between social actors asserting control over nature in accordance with conflicting interests (Bakker 2003; Roberts and Emel 1992; Sheridan 1995; Swyngedouw 2004).

Water scarcity is an example of a concept that can be unraveled in this way. Its definition is typically framed as *physical* scarcity, expressed in terms of supply; for instance:

> Population growth throughout the developing world is increasing pressures on limited water supplies (Gleick et al. 2002: 2).

However, 'supply' is problematic because it fails to either fully investigate or disaggregate scarcity, in terms of *how* water resources become scarce. Although the physical supply of water is important, it cannot be separated from the social relations, which determine how, why and by whom water is used. Thus, when it can be explained by *social*, rather than (entirely) *physical*, factors, scarcity is 'produced', and can be better understood within the hydrosocial cycle (Bakker 2003; Roberts and Emel 1992; Swyngedouw 1997b, 2004). The next section examines the hydrosocial cycle in La Ligua river basin, by examining the ways in which water use and problems are framed by different social actors and how such discourses are mobilized to position favored water management solutions. It also examines the implementation of a physical hydrological assessment and its socioecological implications for water users in the valley.

Water Conflicts in La Ligua River Basin

La Ligua is a transversal Andean valley in the *Norte Chico*[2] in central-northern Chile (Map 4.1). The valley is short and narrow, approximately 20km wide and 100 km long, and characterized by steep sides. The River Ligua rises in the Andean foothills and discharges into the Pacific Ocean. Its source in the low Andes is important geographically, because the river only receives snowmelt in spring, while rivers originating in the high Andes are fed by snowmelt throughout the year. River Ligua therefore experiences a seasonal streamflow, peaking during the spring snowmelt but markedly reduced in the summer. La Ligua basin also contains a shallow and unconfined alluvial aquifer, implying that groundwater and surface water are closely interconnected. The valley is characterized by a semi-arid climate, with an average annual precipitation of 300mm, and a drought approximately every seven years (Gualterio and Curihuinca 2000; Niemeyer and Cereceda 1984).

The principal economic activity in the valley is agriculture, namely fruit cultivation on the valley floor and roaming livestock (cattle, goats) on the valley sides. The valley underwent agrarian reform between the 1960s and 1980s, in an attempt to redistribute large landholdings to landless peasants[3] (Garrido et al. 1988; Kay and Silva 1992; Thiesenhusen 1995). Under the first phase of agrarian reform (1967–1973), land was allocated to peasant collectives, whereby peasants received a small homestead plot and provided labor on the communal landholding. The second phase (until the mid 1980s) undertook 'parceling' projects whereby land 'parcels' (5–20 hectares, depending on land quality) were sold to individual landless peasants. Peasants within parceling projects were also able to buy large areas of rain-fed land on the valley sides.[4] La Ligua is now characterized by a mix of long-established farmers with large estates (100–300 hectares), peasant farmers with land parcels, and newer commercial farmers (with varying sized farms, including parcels purchased from peasants and converted agricultural land on the valley sides).

Under the neoliberal economic program implemented from 1975, commercial export agriculture became a priority for national development, and has led to the expansion and conversion of land to non-traditional export crops in the *Norte Chico* (Gwynne and Meneses 1994; Murray 1997). Since the early 1990s, La Ligua has undergone a shift from annual crops for the domestic market (beans, maize, potatoes, wheat) to permanent fruit plantations for export (avocados, citrus fruits, nuts). The area dedicated to these fruits doubled, from 3619 to 7503 hectares between 1997 and 2002, with avocados by far the dominant crop (INE 1997; ODEPA-CIREN 2002).

---

2   Situated between the arid Atacama desert in northern Chile (*Norte Grande*) and the mediterranean region of Central Chile.

3   Notwithstanding inaccuracy, for the purposes of this chapter the term 'peasant' is used to refer to beneficiaries of agrarian reform with small landholdings, while 'large farmer' is used to denote landholders of over 30 hectares who are engaged in commercial scale production.

4   Carlos Carrera, Agricultural and Livestock Service, Santiago, interview, 28 July 2003.

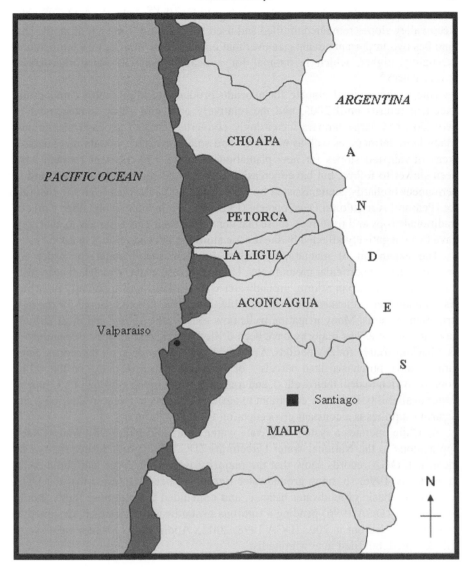

Map 4.1   Location of La Ligua river basin in the *Norte Chico*, Chile (based on IGM 1984)

The expansion of fruit plantations was made possible by two factors: the development of new water and irrigation technology, and the availability of untilled rain-fed land on the valley sides. From the early 1990s, well-drilling, pumps and new irrigation technology, in particular sprinkler or drip systems using PVC pipes, became increasingly mass produced and inexpensive. The new irrigation systems have four important advantages compared with traditional flood irrigation: (i) they can transport water far from the source; (ii) they can irrigate uphill; (iii) they are water-efficient; and (iv) they are labor-saving. In parallel, much rain-fed land on the steep valley slopes remained untilled and used for roaming livestock, if at all. This land has two important advantages over that on the valley floor: (i) the temperature is slightly higher, which is optimal for avocados; and (ii) being rain-fed, it is very cheap.[5]

Due to the optimal climate for avocado production, high export demand and excellent returns until 2005, and the relatively easy and cheap management of avocado trees, large farmers increasingly converted land to avocado plantations. Many large farmers, as well as new large and small investors, bought up extensive areas of rain-fed slopes for new plantations (Figure 4.1). Peasant farmers have been slower to follow, but have increasingly converted some or all of their land to permanent orchards, some assisted by state credit and subsidies from the Institute for [Peasant] Agricultural Development (INDAP). While many maintained a mix of traditional crops and fruit trees, those that converted their entire parcels to avocados have been negatively affected by the drastic slump in avocado prices in 2006.[6]

The expansion of plantations resulted in increased demand for water for irrigation, and in particular groundwater. Unlike surface water, which had been fully allocated after agrarian reform, groundwater was available and apparently plentiful. As groundwater is cleaner and more reliable, it was the favored source for the new irrigation systems. Many irrigation wells now exist in the valley, and most farmers have at least one. The majority have been drilled to irrigate the slope plantations or as a backup source for dry periods. As groundwater is too deep on the slopes, most farmers have purchased land parcels – or even just a strip of one – on the valley floor, on which to drill their well(s), and transport water uphill (through PVC pipes). Water availability is the key constraint to agricultural development in La Ligua, and securing supplies is a constant preoccupation for many farmers.[7]

As Chile operates a system of private water rights under the 1981 Water Code,[8] applications to the National Water Directorate (DGA) for groundwater rights also increased. DGA records show that the majority were from large and commercial farmers. By 1996, so many groundwater rights had been requested that the DGA calculated a basic groundwater balance, and concluded that no more rights should be allocated (DGA 1996) pending a rigorous groundwater assessment, designed in 1998 and carried out in 2002 (DGA 1998, 2002). Applications for new rights were still accepted, but filed in a 'waiting list'.

---

5   Rain-fed land on the valley sides costs approximately US $1430 per hectare, compared with US $14,300–21,500 per hectare for land on the valley floor (2003 prices) (Ariel Zuleta, INDAP Petorca, interview, 11 February 2003).

6   Robinson Sanhueza, INDAP La Ligua, interview, 11 December 2006.

7   AI, large farmer, upper valley, interview, 13 June 2003; PLJ, large farmer, lower valley, interview, 11 September 2003; WJ, large farmer, central valley, interview, 6 June 2003.

8   Chile's 1981 Water Code converted existing water rights (the entitlement to use a certain flow of water under specified conditions) to *private property*. Private water rights are tradable, separate from land, protected by the state and regulated by civil law. The government water agency (DGA) has largely administrative, rather than regulatory, functions. For further details on the application of the Water Code, see Bauer (1997) and Budds (2004).

Figure 4.1   Expansion of avocado plantations in La Ligua

The suspension of water rights did not, however, impede the regularization of historic rights (used since 1976). Since the suspension, regularization became the principal mechanism for acquiring legal groundwater rights. However, it has been used mostly by large farmers, and has been widely abused. Many claims were either dubious or false in terms of the age of the well and the volume of water used 'historically'. Applications were submitted for wells on rain-fed land that had never traditionally been irrigated, some of which were supported by false testimonies from peasants.[9] Although water rights can be purchased from other users, there are too few on the market to be a viable option, and too expensive to be desirable.

Meanwhile, farmers continued to expand plantations and extract groundwater without the corresponding rights. Having no regulatory powers, the DGA was unable to intervene to control illegal extraction. The DGA downplays the scale of illegality in La Ligua,[10] but local evidence suggests that a large number of wells are illegal.[11] Groundwater illegality applies to all types of farmer. However, commercial farmers

9   Substantiated by two lawyers working in the valley (VP, La Ligua, interview, 7 August 2003; SC, Santiago, interview, 17 September 2003), and a review of 50 regularization cases (1995–2003) at La Ligua Civil Court.

10   Humberto Peña, DGA Santiago, interview, 4 August 2003.

11   This is the view of staff from local government agencies and non-governmental organizations, and corroborated by a survey undertaken by INDAP (INDAP 2003). Illegal use is openly admitted by large farmers, often citing the water rights suspension as the impediment to legalization: 'I'm in the same situation as everyone, with some wells registered and others not' (BI, large farmer, central valley, interview, 11 September 2003).

not only have the resources to undertake the regularization process (legitimately or otherwise), but they also attach greater importance to legalizing water rights; partly for their value as private property (capital), partly to access state irrigation subsidies and partly to be able to legally defend their water against potential infractions. Given the barriers faced by peasant farmers to regularize their historic rights – illiteracy and/or unfamiliarity with administrative processes, lack of money to undertake the process – in 2003 INDAP initiated a program to undertake and subsidize the regularization of eligible peasants' wells, and to submit applications for newer wells to the DGA's waiting list.

Contested water scarcity

The agricultural development of the valley has become contentious among different farmers. Agricultural expansion and increasing groundwater use have significantly changed the waterscape, but opinions differ over their likely impacts. These disagreements are based on different perceptions of water scarcity, and are closely aligned with vested interests.

Established large farmers and some peasant farmers, whose farms are on the valley floor – the 'traditional' agricultural heartland – and in the lower valley, resent increasing and uncontrolled groundwater extraction to irrigate the valley slopes, a large proportion of which takes place in the upper valley. They argue that this will exacerbate water scarcity in the whole valley, which, in turn, will jeopardize existing agricultural investments. For example:

> The rain-fed land further up the valley is being irrigated with water that really belongs to the lower valley.[12]

They are particularly worried about the implications for groundwater availability in the next drought (one or more consecutive dry years), since many orchards and livestock were lost in the previous one (1996–1997).

In contrast, farmers who are cultivating the slopes *and* are located in the upper valley, including some peasant communities, generally see the problem as *produced* scarcity. Advocates of this view point out that the cultivated area irrigated with groundwater has expanded significantly since the suspension of water rights, with no apparent impacts. Some assert that the aquifer continually receives snowmelt, while others blame the DGA's failure to undertake an adequate groundwater assessment, and its 'ridiculous' decision to restrict water rights, which is detrimental to both agriculture and economic development. For example:

> Groundwater in Chile is abundant. ... Water scarcity is impossible in Chile. ... Problems with water scarcity are due to infrastructure, not natural scarcity. It is a problem of investment and organization, and a problem that should not exist.[13]

---

12 MJ, large farmer, upper valley, interview, 31 July 2003.
13 AI, large farmer, upper valley, interview, 13 June 2003.

If all agriculture in Chile were using advanced irrigation technology, there would be no water scarcity.[14]

These competing perspectives comprise two dimensions. First, the scarcity discourse is strongly articulated by longer-standing farmers who fear that the large volumes of groundwater extracted by newer farmers could affect their water availability; while the abundance discourse can be linked to newer farmers' need to legalize their groundwater extractions. Second, there is a clear upstream-downstream dimension, whereby farmers in the *lower* valley fear reduced streamflows and declining water tables as a result of increased flows of groundwater used in the *upper* valley.

Discursive solutions

The construction of a reservoir in the upper valley is widely held as the solution to potential irrigation water shortages. A large farmers' consortium has lobbied the government for subsidized infrastructure, arguing that increasing water supply is necessary to fulfill the total irrigation demand for all users, including peasants. This argument is complemented by the observation that excess streamflows during the winter and spring are not presently being stored for irrigation. For instance:

The problem at present is that water is just flowing into the sea, going to waste.[15]

While many peasant farmers also supported a reservoir, not all were convinced that they would benefit. Some even believed that their water would become 'trapped' in the reservoir, and that they would have to pay for it to be released.

Farmers are also divided over proposals to formalize the existing Vigilance Committee,[16] which would then have legal powers to monitor *surface* water allocation and extraction, and mediate conflicts. The debate has centered on whether to constitute one committee for the entire river, or one each for the upper and lower valleys. The divide is clearly territorial, whereby upstream farmers support two committees, whereas those downstream favor one.

The positions are closely tied to two interrelated factors: farmers' spatial location and their interests in terms of securing water availability. First, upstream farmers would benefit from a separate committee under which they would have no obligation to consider the impacts of their water use on the lower valley; while downstream farmers would benefit from the river being managed as one resource, in order to exert some control over use in the upper valley that could affect them. Second, the upper valley has a concentration of established large farmers with original estates and newer commercial farmers with slope plantations who need to secure water. The lower valley is dominated by a majority of small and peasant farmers, who fear that a potential division between the valley would mean that little or no water will flow downstream in dry periods. For example:

---

14 MM, peasant leader, upper valley, interview, 20 June 2003.
15 AI, large farmer, upper valley, interview, 13 June 2003.
16 Water user organization that comprises those irrigation channel users' associations which take surface water from the same river or section of a river (Bauer, 1997).

Those upstream will grab all the water.[17]

The debate over the Vigilance Committee(s) elucidates the politics of water in the valley. The directional flow of water gives upstream users a 'natural' control over water resources against which downstream users are largely powerless. The institutionalization of this biophysical agency by upstream farmers is crucial for the success and continued viability of fruit plantations, and explains why they strive to manipulate the configuration of the valley's Vigilance Committee(s) in their own interests.

The hydrological assessment

The DGA commissioned a comprehensive hydrological assessment for La Ligua in 2002 (DGA 1998, 2002). The assessment calculated the water balance, and created a hydrogeological model to simulate future scenarios of water availability based on different levels of groundwater use (that is, granted water rights and pending applications up to 2003). The basin was divided into 13 sub-sections, and the assessment concluded that most were susceptible to aquifer depletion, due to natural hydrological patterns, but exacerbated by groundwater extraction (DGA 2002). While most sections would recuperate normal volumes following dry periods, three were predicted to undergo permanent depletion, which would also reduce surface water flows. The assessment recommended that some additional water rights could be allocated in most sections, but lower limits were advised for the more vulnerable ones. Based on the assessment, the DGA approved additional groundwater rights applications submitted by late 2003 (DGA 2004a), and declared an aquifer restriction in 2004 (DGA 2004b).[18]

The hydrological assessment contained several limitations, that compromised the accuracy of its results. First, it failed to adequately acknowledge the limitations of the model, primarily the degree to which it accurately reproduced the basin's water system. Second, it omitted a both a sensitivity analysis (which assesses the extent to which variations in the input data affect the simulated results), and margins of error for the results (the acceptable deviation from the specific value), so the accuracy of the simulations is unknown. Third, the quality of some input data was deficient. The model only used field streamflow measurements for the main river, while those for the ungauged tributaries were estimated. As these data are important inputs, any error will reverberate through the calculation of flows in the entire model. Data for agriculture and irrigation were also outdated, so crucial recent increases were not fully represented. Fourth, the model used groundwater rights as a proxy for water actually used. This is problematic because some farmers have more rights than they use, while others use water without the corresponding rights. The omission of the large amount of illegal groundwater use is particularly significant, because this is approximately *double* that of the rights considered.[19] Furthermore, the application

---

17 RA, peasant leader, central valley, interview, 19 June 2003.

18 A restriction means that no new groundwater rights can be allocated.

19 Illegal use (13,859 liters per second, compared with 7508 liters per second used legally) was estimated from the INDAP survey (INDAP 2003).

of a 'coefficient of effective use' to estimate the proportion of water actually used probably underestimates usage (DGA 2002).[20]

Nevertheless, the results of the assessment were adopted as the basis of policy, with no apparent consideration of the limitations (DGA 2004a). The results were presented as definite values, and a single figure of 1547 liters per second was adopted from one scenario to represent the amount of water available for allocation, despite the inaccuracies permeating the assessment. However, this figure alone was only sufficient to grant new water rights requested by October 1996. Other factors were then included to increase the flow of available water, although these were *not* included in the modeling exercise. These comprise: (i) water that is not used by crops and returns to the source ('return flow'), (over)estimated at 80 per cent; (ii) leakage from water supply; and (iii) an additional allocation of 'provisional' rights.[21] The new figure, 5542 liters per second, was sufficient to cover *almost all* of the water rights requested by November 2003, although the model had only included granted water rights and new applications until March 2001.

Although the situation in La Ligua was characterized by increasing groundwater exploitation and mounting social tension, the DGA commissioned a purely *physical* water resources assessment. The only data on water use were groundwater rights, which neither reflected irrigation practices (such as occasional well use in dry periods) nor the widespread illegality. Therefore, the assessment only considered physical flows of water and omitted qualitative factors such as patterns of use between different users. It also neglected local knowledge, from farmers, government institutions or non-governmental organizations, such as fluctuations in groundwater levels, location of springs or changes over time.

As required by Chilean law, the new groundwater rights were allocated by order of submission (DGA 2004a). However, this policy mechanism is *socially* problematic in three ways. First, the assessment failed to disaggregate the roles of different users in potential groundwater overexploitation. Second, it ignores differential access to the water rights application system based on socio-economic status. Third, it does not consider the social implications of the policy decision. Furthermore, it is also *ecologically* problematic. On the one hand, it pays no attention to the location of groundwater, so new rights can be granted in aquifer sections flagged as already vulnerable. Indeed, many applications were for the upper valley, which could affect downstream users. On the other hand, the factoring-in of return flows for new allocation reverses the order in which users receive water. If many new rights are allocated upstream of existing rights –which *is* likely– new users will receive water first, and existing users will rely on their return flow. Given the spatial concentrations of large farmers in the upper valley, and peasants in the lower valley, the *socioecological* implications of this policy have the potential to significantly impact downstream water users.

Legislative modifications to the Water Code in 2005 included a simpler mechanism for regularizing small wells (up to two liters per second) built by June 2004. The mechanism was initially valid for six months, but later extended to one

---

20 For irrigation, the coefficient assumes that only 20 per cent of water is actually used.

21 Temporary rights that are allocated when permanent rights are exhausted, and are cancelled after five years if negative impacts on the aquifer are detected.

year. It has been welcome for peasant farmers, and has enabled many to gain legal water rights for wells that would otherwise have remained illegal. However, in many cases the process was impeded by peasants requiring assistance with the application, and by outdated land tenure documents.[22] The mechanism was also widely exploited by large farmers, some of whom submitted over 100 wells.[23]

Contested Scarcity and Hydrosocial Change

In La Ligua, water resources, scarcity and illegality are all unevenly distributed between large and peasant farmers. While large farmers have the awareness and resources to access water rights and lobby government bodies for solutions, peasants have fewer (legal) water rights and the exhaustion of new groundwater rights jeopardizes their irrigation security. Already suffering from low avocado prices, potential losses during the next drought, and their livelihood implications, may be severe. Wider temporal-spatial processes relating to land and water have influenced these social relations of access to water, and the physical waterscape. Historically, agrarian reform and export agricultural policies have largely exacerbated existing inequalities and made little improvement to rural poverty (Kay and Silva 1992). In La Ligua, reformed land has been regained by large farmers and converted to fruit plantations, while large volumes of water are pumped from the valley floor to the slopes, with little attention to the water security of existing users.

The interconnectedness and directional flow of water facilitates this hydrosocial change. Unlike land, water allocations cannot be alienated, and can be difficult to protect. Instead, upstream users gain a 'natural' advantage, by being able to access water before it flows downstream. This is reflected in attempts by upstream farmers to form a separate Vigilance Committee that would exploit this biophysical agency and absolve them of responsibility for downstream users. The debate over the formation of the Vigilance Committee(s) transgressed into a power struggle between farmers who organized themselves on a predominantly *spatial*, rather than a *class*, basis to support the option that best coincided with their interests; demonstrating the ability of water's agency to shape the social relations around it.

The geography of water in the valley also shapes powerful discourses used to exert social control over water. While physical hydrological conditions can produce water scarcity, the discourses of 'scarcity' and 'abundance' are used to mobilize different farmers' interests. Thus, upstream farmers argue that the aquifer restriction reflects arbitrary and unnecessary government bureaucracy, and lobby for the reversal of the restriction; while downstream farmers criticize the over-allocation of groundwater rights and petition the state to sanction illegal extraction. Farmers favoring a reservoir relate water scarcity to seasonal fluctuations in river flow or 'wastage' of water into the ocean. These arguments blur physical and produced scarcity by framing the problem in terms of hydrogeological conditions and water's materiality, which justifies a supply-led solution, while ignoring patterns of water use, namely increased irrigation of the valley slopes. Moreover, framing

---

22 Carmen Cancino, INDAP, personal communication, 18 April 2005.
23 Leonardo Olivares, Provincial Government of Petorca, interview, 12 December 2006.

the problem in this way detracts attention from their *own* role in causing it. In this way, they discursively construct particular 'narratives' (Cronon 1992) or 'storylines' (Hajer 1995) to frame the water situation in a way that tacitly supports their own particular motives and desired ends. These framings are then mobilized to position these particular explanations as dominant, which will garner political support and acceptance that, in turn, will deliver benefits to those social actors who promoted their own situated readings of the problem (see also Boykoff, this volume).

Such discourses of 'scarcity' and 'crisis' are generalized to all farmers in the valley, especially peasants, in order to garner support for favored solutions. The need for a reservoir is promoted on the basis that it will foster development of the valley, with benefits to *all* farmers. Peasants have been co-opted into supporting a reservoir, but few have questioned *why* that solution is being promoted so forcefully. Indeed, it could be argued that it is precisely the large farmers undertaking agricultural expansion who are potentially exacerbating water scarcity and jeopardizing water security, in particular for peasants; yet it is also they who are supporting supply-driven solutions to this situation, but in the name of all farmers – and especially peasants – even though they themselves are likely to be the principal beneficiaries. Such discourses are only possible due to the interconnected, flowing and inalienable nature of water, which obscures the social relations of its control.

The integration of the scales at which both social and ecological processes occur gives an insight into the complexity of the situation in the valley. In particular, there is a mismatch between the scales at which processes are observed, situations are caused and remedial measures are applied. The hydrological assessment oversimplified hydrosocial processes by focusing exclusively on aquifer depletion as a result of increased groundwater extraction, and neglecting the potential effects of the relocation of groundwater use within the basin, especially in the upper valley. By only considering the hydrological cycle, the assessment restricted its analysis to the basin scale and thus narrowly examined local processes, namely groundwater extractions. As a result, it omitted underlying socio-political processes that shaped hydrosocial change, such as the Water Code and export agricultural policies, yet which operate beyond the space-time boundaries of the basin. Similarly, the policy response was homogenous in terms of its uniform application to both water users and the river basin, with potentially uneven socioecological outcomes. It was precisely the dualistic view that separates people from water, as well as the conceptualization of water as purely material, that permitted policies that produced differential access to water.

The choice of a *physical* hydrological assessment shaped how the water situation and its solutions were framed in the ensuing policy formulation. Focusing exclusively on the materiality of water had several implications. Considering water in desocialized terms obfuscated the fact that peasant farmers have not been largely responsible for any potential aquifer depletion, yet stand to lose most from the restriction. It also enabled the situation to be framed as an *environmental* issue rather than a *social* or *political* one. This itself legitimated a physical approach, positioned as a technical and accurate assessment that would reliably inform water resources decisions. The DGA presented the scientific assessment as the only legitimate knowledge about water resources, but manipulated the results by ignoring its limitations, incorporating other calculations in order to satisfy the majority of the demand (as it was under pressure to do, mainly from large farmers), and using the narrative of 'depletion' to impose a groundwater restriction. This also allowed the DGA to focus on the hydrological dimension (the aquifer restriction), while leaving the social dimension (the allocation of new groundwater rights) to be determined by administrative processes under the Water Code, thus deftly sidestepping the thorniest element of the situation. The decision was top-down and non-negotiable, framed as administrative and neutral, yet powerful in that it carried significant socioecological implications.

The Dialectics of Power and the Waterscape

Through the case study of La Ligua river basin, this chapter sought to analyze the material, sociopolitical and discursive elements of a real water conflict, based on the reconceputalisation of water from a 'resource' to a 'socio-nature'. In turn, this highlighted several important and nuanced dimensions. The chapter explored how water is deeply politicized by the different actors in La Ligua, as different types of farmer struggled to secure irrigation water to meet their needs, and vied to influence modes of water management in line with their particular vested interests. However, rather than considering water as merely a static object over which power is exerted, the dialectical relationship between social power and hydrosocial change also illustrated how the materiality of nature – its biophysical properties and agency – configured the social relations of control over it. In particular, due to water's dynamic and directional flow, upstream users enjoyed a 'natural' priority over access and farmers organized on a spatial, rather than a class, basis, with each group socially constructing groundwater as scarce or abundant, according to their location in the valley and agricultural interests.

The chapter also explored the implementation of a hydrological assessment, the results of which formed the basis of water policy for the basin. Although framed as scientific, by failing to consider its technical limitations, and by focusing on purely physical and quantitative processes and data, the accuracy of the results of the model simulations were questionable. Moreover, the assessment's exclusive treatment of water as a material resource circulating within the physical environment failed to represent, and, indeed, served to subjugate, the *political* conflict over water. Similarly, based on technical environmental science, the assessment was positioned as *independent*, the water situation as *environmental*, and the subsequent policy processes as *administrative*. These discourses of neutrality – that are particularly powerful because they seek to deny their own situated nature – became important vehicles for enabling socioecological inequalities in their practical implementation, through the implementation of the desocialized water resources assessment and the resulting depoliticized allocation of water rights.

The resulting decisions had the potential to reduce the water security of the poorest group of farmers in the valley, who were least responsible for the overexploitation of groundwater to which the assessment responded. This outcome was not only exacerbated by failing to disaggregate water (ab)use among different farmers, but also by treating the basin as a homogenous scale, despite the upstream-downstream dynamic. In particular, the eventual allocation of new groundwater rights produced potentially uneven socioecological outcomes, by both failing to consider access to rights, and by reorganizing the spatial use of water in the basin. Given that the lower valley is populated by a large number of peasant communities, the 'social' and 'ecological' dimensions of the outcomes are likely to reinforce each other. This implies that peasant farmers will become the worst off, not just because they are poor, but also because they are downstream.

The approach to assessing water resources in La Ligua clearly prioritized the hydrological cycle, by privileging the estimation of physical water flows using scientific and quantitative methods, as a means of both producing knowledge and determining water rights allocation. In this way, it entirely dismissed the *hydrosocial* context within which water also flowed. As such, it ignored the conflict over water that was the very reason for undertaking the assessment; it neither paid attention to qualitative factors (such as patterns of use between different users) nor the institutional framework within which use, access and control of water was embedded (the Water Code); and it neglected local knowledge, from farmers, government institutions or non-governmental organizations. In contrast, a *hydrosocial* assessment would have given equal importance to sociopolitical factors as meteorological and streamflow data, and would also have considered local knowledge and participation in the research process, thus overturning the dominant view that only formal scientific knowledge is valid. This would not only have produced a more comprehensive, legitimate and democratic water resources assessment, but would also have reduced the potential for top-down decisions to be made. It would, however, have significantly challenged existing dominant traditions and power structures.

Acknowledgements

The author would like to thank the editors for all their work in compiling this volume; her doctoral supervisor, Erik Swyngedouw (Manchester), and co-supervisor, Mark New (Oxford); the Natural Resources and Infrastructure Division at ECLAC in Santiago, which supported this fieldwork through an internship; Carmen Cancino at INDAP in Santiago, and Ariel Zuleta, Robinson Sanhueza and colleagues at INDAP's local office in La Ligua; the many others, especially farmers, who gave interviews for this research and assisted in other ways; Klaus Hellgardt (Imperial College) for help with understanding hydrological modeling and drawing the map; and financial support through an ESRC/NERC PhD studentship.

References

Bakker, K. (2003), *An Uncooperative Commodity: Privatizing Water in England and Wales* (Oxford: Oxford University Press).

Bauer, C. (1997), 'Bringing Water Markets Down to Earth: The Political Economy of Water Rights in Chile, 1976–1995', *World Development*, 25: 5, 639–56.

Blaikie, P. (1985), *The Political Economy of Soil Erosion in Developing Countries* (London: Longman).

—— (1995), 'Changing Environments or Changing Views? A Political Ecology for Developing Countries', *Geography*, 80: 3, 203–14.

—— (2001), 'Social Nature and Environmental Policy in the South: Views from Verandah and Veld', in Castree and Braun (eds).

Braun, B. and Wainwright, J. (2001), 'Nature, Poststructuralism, and Politics', in Castree and Braun (eds).

Bryant, R. and Bailey, S. (1997), *Third World Political Ecology* (London: Routledge).

Budds, J. (2004), 'Power, Nature and Neoliberalism: The Political Ecology of Water in Chile', *Singapore Journal of Tropical Geography*, 25: 3, 322–342.

Castree, N. (2001a), 'Marxism, Capitalism, and the Production of Nature', in Castree and Braun (eds.).

—— (2001b), 'Socializing Nature: Theory, Practice, and Politics', in Castree and Braun (eds).

Castree, N. and Braun, B. (eds) (2001), *Social Nature: Theory, Practice, and Politics* (Oxford: Blackwell).

Cronon, W. (1992), 'A Place for Stories: Nature, History, and Narrative', *Journal of American History* 78: 4, 1347–1376.

Demeritt, D. (1998), 'Science, Social Constructivism and Nature', in Braun, B. and Castree, N. (eds) *Remaking Reality: Nature at the Millennium* (London: Routledge).

—— (2001), 'Being Constructive about Nature', in Castree and Braun (eds).

DGA (1996), 'Determinación de la disponibilidad de recursos hídricos para constituir nuevos derechos de aprovechamiento de aguas subterráneas en el sector del acuífero del valle de La Ligua, Provincia de Petorca, V Región' [Assessment of the availability of water resources to grant new groundwater rights in the aquifer of La Ligua basin, Petorca Province, Fifth Region], Technical Brief No. 13 (Santiago: Dirección General de Aguas [National Water Directorate]).

—— (1998), 'Análisis y evaluación de los recursos hídricos de las cuencas de los ríos Petorca y Ligua' [Analysis and evaluation of water resources in Petorca and La Ligua river basins], Technical Report (Santiago: Dirección General de Aguas [National Water Directorate]).

—— (2002), 'Evaluación de los recursos hídricos, Cuenca del Río Ligua, V Región' [Evaluation of water resources, La Ligua River Basin, Fifth Region], Technical Report No. 80 (Santiago: Dirección General de Aguas [National Water Directorate]).

—— (2004a), 'Actualización de la evaluación de los recursos hídricos subterráneos acuífero cuenca del río La Ligua, Vª Región' [Update and evaluation of groundwater resources in the aquifer of La Ligua river basin, Fifth Region], Technical Report No. 166 (Santiago: Dirección General de Aguas [National Water Directorate]).

—— (2004b), 'Área de Restricción para un sector del acuífero del valle del río La Ligua' [Restriction area for a section of the aquifer of La Ligua river basin], Technical Report No. 54 (Santiago: Dirección General de Aguas [National Water Directorate]).

Escobar, A. (1999), 'After Nature: Steps to an Antiessentialist Political Ecology', *Current Anthropology* 40: 1, 1–30.

Forsyth, T. (2003), *Critical Political Ecology: The Politics of Environmental Science* (London: Routledge).

Foucault, M. (2002), 'Governmentality', in J. Faubion (ed.) *Power / Michel Foucault* (London: Penguin).

Garrido, J., Guerrero, C. and Valdés, M. (1988), *Historia de la Reforma Agraria en Chile* [History of Agrarian Reform in Chile] (Santiago: Editorial Universitaria).

Gleick, P., Wolff, G., Chalecki, E. and Reyes, R. (2002), *The New Economy of Water: the Risks and Benefits of Globalization and Privatization of Fresh Water* (Oakland, California: Pacific Institute).

Gualterio, H. and Curihuinca, J. (2000), *Zonificación Agroclimática: V Región* [Agro-climatic Zoning: Fifth Region] (Santiago: Dirección Meteorológica Nacional [National Meteorological Directorate]).

Gwynne, R. and Meneses, C. (eds) (1994), 'Climate Change and Sustainable Development in the Norte Chico, Chile: Land, Water and the Commercialisation of Agriculture', Occasional Publication/Research Report (Oxford: University of Birmingham and University of Oxford).

Hajer, M. (1995), *The Politics of Environmental Discourse: Ecological Modernization and the Policy Process* (Oxford: Clarendon Press).

Haraway, D. (1991), *Simians, Cyborgs and Women: the Reinvention of Nature* (London: Routledge).

Harvey, D. (1996), *Justice, Nature and the Geography of Difference* (Oxford: Blackwell).

IGM (1984), *Mapa Hidrográfico de Chile* [Map of Hydrological Basins of Chile] (Santiago: Instituto Geográfico Militar [Military Geographic Institute]).

INDAP (2003), *Catastro de obras de captación de aguas subterráneas de propiedad de pequeños productores agrícolas cuyos derechos de aprovechamiento no se encuentren inscritos y/o regularizados en la cuenca hidrográfica del Río La Ligua* [Survey of infrastructure for extracting groundwater resources and the legal status of the water rights among small agricultural producers in La Ligua River Basin] (database) (Santiago: Instituto de Desarrollo Agropecuario [Institute for Peasant Agricultural Development]).

INE (1997), *VI Censo Nacional Agropecuario* [Sixth National Agricultural Census] (Santiago: Instituto Nacional de Estadísticas [National Statistics Institute]).

Johnston, R., Gregory, D., Pratt, G., Watts, M. and Smith, D. (eds) (2000), *The Dictionary of Human Geography*, 4th edition (Oxford: Blackwell).

Kay, C. and Silva, P. (1992), *Development and Social Change in the Chilean Countryside: From the Pre-Land Reform Period to the Democratic Transition* (Amsterdam: CEDLA).

Latour, B. (1993), *We Have Never Been Modern* (London: Longman).

Mansfield, B. (2005), 'Beyond Rescaling: Reintegrating the 'National' as a Dimension of Scalar Relations', *Progress in Human Geography* 29: 4, 458–473.

Marston, S., Jones, J.P. and Woodward, K. (2005), 'Human Geography without Scale', *Transactions of the Institute of British Geographers* 30: 4, 416–432.

Murray, W. (1997), 'Competitive Global Fruit Export Markets: Marketing Intermediaries and Impacts on Small-scale Growers in Chile', *Bulletin of Latin American Research* 16: 1, 43–55.

Niemeyer, H. and Cereceda, P. (1984), *Geografía de Chile: Hidrología* [The Geography of Chile: Hydrology], Volume VIII (Santiago: Instituto Geográfico Militar [Military Geographic Institute]).

O'Riordan, T. (1999), 'Ecocentrism and Technocentrism', in Smith, M. (ed.) *Thinking Through the Environment: A Reader* (London: Routledge).

ODEPA/CIREN (2002), *Catastro Frutícola – V Región* [Fruit Production Survey – Fifth Region] (Santiago: Oficina de Estudios y Políticas Agrárias [Agrarian Research and Policy Office] and Centro de Información de Recursos Naturales [Natural Resources Information Center]).

Paulson, S. (2003), 'Gendered Practices and Landscapes in the Andes: The Shape of Asymmetrical Exchanges', *Human Organization* 62: 3, 242–254.

Paulson, S. and Gezon, L. (eds) (2005), *Political Ecology Across Spaces, Scales, and Social Groups* (Piscataway: Rutgers University Press).

Paulson, S., Gezon, L. and Watts, M. (2003), 'Locating the Political in Political Ecology: An Introduction', *Human Organization* 62: 3, 205–217.

Robbins, P. (2004), *Political Ecology: A Critical Introduction* (Oxford: Blackwell).

Roberts, R. and J. Emel (1992), 'Uneven Development and the Tragedy of the Commons: Competing Images for Nature-society Analysis', *Economic Geography* 68: 3, 249–71.

Sheridan, T. (1995), 'Arizona: The Political Ecology of a Desert State', *Journal of Political Ecology* 2, 41–57.

Swyngedouw, E. (1997a), 'Neither Global nor Local: "Globalization" and the Politics of Scale', in Cox, K. (ed.) *Spaces of Globalization: Reasserting the Power of the Local* (New York: Guildford).

Swyngedouw, E. (1997b), 'Power, Nature and the City: The Conquest of Water and the Political Ecology of Urbanization in Guayaquil, Ecuador: 1880–1990', *Environment and Planning A* 29: 2, 311–332.

—— (1999), 'Modernity and Hybridity: Nature, Regeneracionismo, and the Production of the Spanish Waterscape, 1890–1930', *Annals of the Association of American Geographers* 89: 3, 443–465.

—— (2004), *Social Power and the Urbanization of Water: Flows of Power* (Oxford: Oxford University Press).

Thiesenhusen, W. (1995), *Broken Promises: Agrarian Reform and the Latin American Campesino* (Boulder: Westview).

Walker, P. (2005), 'Political Ecology: Where is the Ecology', *Progress in Human Geography* 29: 1, 73–82.

Whatmore, S. (2002), *Hybrid Geographies: Natures, Cultures and Spaces* (London: Sage).

Zimmerer, K. and Bassett, T. (eds) (2003), *Political Ecology: An Integrative Approach to Geography and Environment-Development Studies* (New York: Guilford).

PART 2
Conflicting and Shifting Environmental Knowledges, Livelihoods, and Power

Chapter 5

'Environmentality' in Rajasthan's Groundwater Sector: Divergent Environmental Knowledges and Subjectivities

Trevor L. Birkenholtz

Introduction

This chapter draws on Arun Agrawal's (2005) conceptual framework of 'environmentality' to examine the decentralization of environmental governance in India. The more precise goal is to exemplify how environmentally aware subjects – i.e. 'environmental subjects' – are actively produced by the state, how the state and its subjects understand each other, and what this means for the future creation of formal groundwater management institutions in Rajasthan, India. A combination of environmentalism and Foucault's governmentality (1991), environmentality is a method for examining the 'creation of governmentalized localities, institutional politics within regulatory communities, and the making of environmental subjects' (Agrawal 2005: 20). It is a way of understanding the effects and functions of the decentralization of environmental governance by looking to 'knowledges, politics, institutions, and subjectivities that come to be linked together with the emergence of the environment as a domain that requires regulation and protection' (Agrawal 2005: 226).

This interpretive framework informs the present work, conceptually and methodologically, in four distinct ways. First, it is concerned with the ways in which the environment is discursively framed as an object of concern by the state and, in this instance as something that needs to be conserved and protected. Second, it examines the ways in which this new environmental awareness relies on the interaction of different forms of knowledge but especially the deployment of that knowledge by the state that is commonly defined both by regulator and regulated as 'expert' and on which the authority of conservation rests. Exploring this confluence of topics opens up a space in which to understand both how people understand or 'see the state' (Corbridge et al. 2005) and what this might mean for successful regulation. Third, environmentality shows how these ideas, and the rules needed to promote them, are most effective when their production and enforcement is decentralized. And fourth, it highlights that the legitimation of these ideas cannot be imposed but must be instilled, in accordance with the norm of decentralization, through 'governmentalized localities'; this depends on the simultaneous production of environmental subjects, who will willingly perform the tasks of monitoring and enforcement. These concrete concerns can be expressed in the form of three distinct questions: 1) how and why does the state produce environmental subjects and what are the tensions in this process?; 2) how does looking to divergent state and local environmental knowledges inform this process?; and 3) what does this mean for the future efficacy of decentralized environmental regulation? The future of the regulation of Rajasthan's groundwater resources, at least in the short-term, depends upon bridging the gap between state and lay environmental knowledge and how these two articulate in particular mutual understandings of each other and of the problems and solutions to groundwater overdraft.

In an attempt to better understand the active production of environmental subjects by the state and the implications for groundwater governance, I examine the Indian state's recent efforts to create decentralized regulation in the proposed 'Rajasthan Groundwater Rational Use and Management Act of 2005' (hereafter the Act). The Act has two key provisions. First, it sets up a new hierarchical Groundwater Authority to implement the Act's regulations, whose future effectiveness rests (secondly) on its aim to create a sense of environmental awareness in groundwater among peasant farmers. Given that the Groundwater Authority has yet to come into being and begin its 'work', my analysis is limited to 1) an examination of the provisions for the organization of the Authority, which are top-down and provide limited opportunities for the mutual exchange, between the Authority and farmers, of understandings and knowledge of groundwater practices and management institutions, and 2) the actions of current state agencies in Rajasthan that work with farmers on groundwater issues and, how and why farmers relate negatively to them.

In this chapter, I analyze the structure and character of the proposed Act to evaluate the prospects for the creation of successful groundwater regulation. First, I demonstrate that the proposed regulation is hierarchical in that it does not provide a democratic mechanism. Second, by examining the proposed groundwater regulation, I find that current regulatory efforts do not incorporate locally existing knowledge, practices or institutions of groundwater use, but instead plan to alter the way people currently see groundwater. Third, the likely implications of this failure are highlighted by divergences between local and state groundwater knowledge in deep tubewell construction. So while the Rajasthan state is indeed trying to foster a widely held sense of environmentalism in water resources, as both a precondition of decentralized environmental governance and as a critical element of self-regulating environmental people, the effect is likely to be compromised, due in part to the rift between local and state groundwater knowledge. The other part of the problem is that these efforts seem to be part of an overall strategy to decentralize state power, rather than to decentralize governance, and this could work counter to the creation of environmental subjects and 'successful' groundwater governance.

Following Agrawal, successful natural resource regulation – whether to the actual benefit of people or not – is dependent both on the state enrolling people into its environmental cause and creating a non-hierarchical and democratic means of ensuring local input (whether real and imagined) into the policy making and enforcing process. Similarly, Martin (2003) draws on a analogous approach to identify the rifts between local and state (or 'expert') knowledge in Indian Forestry, leading him to call for a more 'integrative approach to natural resource management that does not isolate community participation from its wider contexts (ibid: 67, see also Scoones and Thompson 1994; Robbins 2000)'. In Britain, Morris (2006) points to the divergences between farmer and state understandings of more sustainable farming practices to show how these might adversely affect policy outcomes. Social science research that examines the relationship between, and the ways that tensions emerge, among local and state forms of expertise, is useful and relevant therefore, in predicting the potential impact that these interactions have on environmental policymaking. This present work evaluates the prospects for successful groundwater regulation along the lines of successful forest regulation outlined by Agrawal and others. It also offers an opportunity to advance our understanding of the processes of decentralized environmental governance through *active* state environmental subject making and the contentious geographies they make.

The chapter first outlines the four aspects of environmentality outlined above: 1) framing the environment as something in need of protection; 2) which relies on the deployment of expert knowledge; 3) to institutions of decentralized governance that result in the creation of 'governmentalized localities' and 4) the making of environmental subjects. In the second section, I lay out recent attempts by the State of Rajasthan to draft new groundwater legislation, which has yet to be fully implemented. In the third section, I advance the lessons from Agrawal, by applying the concepts of environmentality to the problem of groundwater conservation and regulation in Rajasthan. The research for the present paper was performed in India and draws on other research undertaken there. However, the lessons exemplified about the decentralization of environmental governance are more general in character.

Environmentality, Governance, and Technologies of Rule

In his book *Environmentality* (2005), Agrawal analyzed how villagers in the Kumaon region of India's Himalayan state of Uttaranchal, went from setting forest fires in the early 1920s, protesting British colonial efforts of forest enclosure and extractive conservation, to being active environmental stewards in the 1990s. The key to this transformation, he argued, was the decentralization of environmental governance through the environmental subjectification of forest users. Agrawal identified three aspects of decentralized environmental governance: 1) 'the redefinition of political and administrative links between the state and localities', which he termed 'governmentalized localities'; 2) 'the realignment of institutional and social relationships within local communities, and 3) the emergence of a more widespread concern with the environment and the making of environmental subjects' (Agrawal 2005: 89). In Rajasthani groundwater governance, these relationships are still in a process of 'becoming'. Since the Act has yet to come into force, it is not yet possible to follow through on the second point above; i.e. 'the realignment of institutional and social relationships within local communities'. Consequently, I examine points one and three by using examples from his text.

'Governmentalized localities'

The British East India Company took control of Kumaon in 1815. At this time, the forests were thought to be inexhaustible. Initially they were not heavily used and 'for years nothing was done to protect forests in any way' (Bailey 1924, cited in Agrawal 2005: 68). But in the latter half of the 19th century, Indian railway construction demanded increasing amounts of wood for railroad ties, which increased the value of the forests. This led to rampant speculation as the government leased out vast areas to private forestry contractors at negligible rents and without oversight. Many clear-cut their leased areas and, with disregard, took only the largest timbers. This prompted the first official survey in 1869 that concluded 'the forests have been worked to desolation', even though the contract system was abolished over ten years previously in 1858 (Pearson 1869, cited in Agrawal 2005: 68).

The drastic deterioration in the condition of forests led to the Forest Act of 1865, which brought the most valuable areas under state control. This Act marked the beginning of scientific forestry in India grounded in formal expertise. Foresters defined what a 'forest' was and what it was not, how it should be managed, and what activities should be allowed or excluded. The department also measured tree girth, took surveys, and used statistics to quantify the forest growth, determine sustained yields and inform reforestation objectives. The increased reliance on statistics to manage forests created a 'new generation of foresters who had faith and confidence in the technologies of government of which they were a part' (Agrawal 2005: 60).

This led to increasing disregard for peasant forest use and management, which was viewed as 'unscientific'. The numbers were used as 'automatic pilots' in decision making, with the effect of depoliticizing and justifying environmental government, including the exclusion of peasant activities and peasants themselves (Rose 1999: 205). But displaced and otherwise affected villagers rejected this proposition, leading to increased peasant unrest and protests. The Forest Act of 1878 added to this tension by demarcating forests and dividing them into reserved forests, protected forests, and village forests. This increased the area under state control dramatically. As a result, by 1916 the state had 3,000 square miles of reserved forests (up from 200 in 1911) (Agrawal 2005: 72). The forest department now had the authority to restrict and regulate all activities in reserved forests, but it did not have adequate staff to enforce the new regulations. This had two effects.

First, it led to yet more peasant protests, including the burning of 200,000 acres of forest in 1916 (ibid: 3). Second, the scarcity of regulatory officials, the economic and political costs and inefficiencies of centralized regulation, and the continuing peasant protests prompted a new regulatory approach embodied in the Forest Council Rules of 1931. The rules created a system of decentralized forestry management by bringing nearly 2,000 square miles under the shared regulatory jurisdiction of the forest department, the revenue department and village residents (even though forest officials were still skeptical that villagers could manage their forests) (ibid: 83). Enrolling villagers into the system of forest conservation (ultimately for maximum sustained yield and profit) required changing their attitudes towards Forest Department policies, while bringing them into the fold of regulation. The solution was found through what Agrawal termed 'governmentalized localities' (ibid: 101).

Governmentalized localities were both a strategy and an outcome of these processes. The Forest Council Rules resulted in more than 3,500 such councils today (one-third of all villages in Kumaon) and were the precursor of today's Joint Forest Management (JFM) (ibid: 79, 119). They represented and continue to represent a shift in environmental governance and a new technology of government that is less costly, both economically and politically. But they are also the outcome of peasant resistance to centralized, non-representative government. The forest councils radically transformed the regulatory landscape, but instead of creating localized mirror images of the state, they relied on existing forms of cooperation and institutions, and existing local governments (*panchayats*). Therefore, the Rules created 'governmentalized localities', which localized some authority and encouraged (and depended for their success on) 'the willing participation of those subject to rule and rules' (ibid: 125). In other words, it depended on the third aspect of Agrawal's decentralized environmental governance – the state inscription of environmental awareness and the making of environmental subjects. It needed to produce a particular kind of local subject to carry forward its strategies, but the subjects the process produced were not always predictable and not everyone became an environmentalist. It is to environmental subjects that I now turn.

Environmental subjects

For those who have become environmental subjects, the 'environment constitutes … a conceptual category that organizes some of their thinking; it is also a domain in conscious relation to which they perform some of their actions' (Agrawal 2005: 165). The human turns himself or herself into a subject 'by following certain practices and modes of thought' (Agrawal 2005: 221). The process of subject formation depends, therefore, on the intersection of power/knowledge, but is also dependent upon prior political, economic and social positioning. This is why not all people subjected to the same conservation message will become environmentally aware. The questions are: how does the state make environmental subjects?; is this process one-sided and antagonistic, or mutually dependent?; and why do some come to care for the environment while others do not?

Agrawal contends that active participants in the regulatory process, such as those who make rules, monitor forests or allocate resources, gain an understanding of the natural resources that are in their charge and see the effects of human use, both positive and negative as they view them from their situated position within this process. Therefore, regulations cultivated through governmentalized localities and mutual exchange and obligation are technologies of state power, based on the mobilization of knowledge, which encourages subjects to define themselves in relation to it. Again following Agrawal (2005), 'regulatory rule, creates awareness and knowledge through direct participation in the various elements and stages of regulations' (163). So too, the subject does not simply emerge and exist independently of historical, political and social conditions. But, again following Agrawal, the subject cannot be understood simply as a product of particular social differences such as those of caste, class or gender. He argues that it is practice that differentiates various kinds of subjects. Therefore, while social categories are important it is more than looking simply to them. We need to examine individual practices of articulating with these new regulatory regimes (such as rule making and enforcement), which guide particular understandings and perceptions of environmental problems and their solutions (ibid: 197).

Over time, communities assume a sense of ownership over their resources as they define and carry out rules and regulations. In what Agrawal termed 'imagined autonomy', villagers used the state language of regulation and protection in pursuit of management goals that they perceived to be their own, while in fact, they came from the state. Imagined autonomy actually derived from performing the conservation practices encouraged by the state. This produced a sense of environmentalist identity that was crucial for successful decentralized environmental governance. It also produced a sense of democratic rule making and enforcement that was both real and imagined. It was real in the sense that some of the peoples' goals were met (through protest), such as the granting of fodder harvests. But it was imagined in the sense that the environmentalist subjectivities that people took to be of their own making, actually came from the state. It is this real and imagined democratic autonomy that made forest regulation successful. And while Agrawal contended that this form of governance has led to mutually beneficial conservation outcomes, it is also possible that it could be a form of government that allows more effective control over local populations and their resources.

The twin concepts of governmentalized localities and environmental subjectivity are critical to the recent Rajasthan state efforts to regulate groundwater use. Thinking through what Agrawal's conceptual framework and recent regulatory reforms respectively mean for the water sector in India is particularly timely. This is because groundwater is a contentious issue in the country due to its importance for irrigation and domestic purposes as well as the object of newly proposed regulation.

Regulating Groundwater in Rajasthan:
The Proposed Groundwater Bill and Making Environmental Subjects

Study area – Jaipur district, Rajasthan

Rajasthan is divided into 32 districts and each district is divided into numerous blocks or *tehsils* (there are 237 in all). Research for the present study took place in the summer of 2005. I surveyed over 150 farmers in six villages of Bassi Tehsil, around 60 kilometers east of Rajasthan's capitol city, Jaipur (see Map 5.1). Then, over the next several months I followed up on these surveys with in-depth interviews of over 78 farmers, multiple government technocrats, and several tubewell drilling firms. Chosen for their social and ecological diversity, the villages comprise a highly stratified social environment of low and high caste Hindus, small, medium and large landholders, and moderately to rapidly declining groundwater levels in their respective vicinities. Even though we should not take social categories, such as caste and class, to assume difference, they are still important considerations (Jeffrey 2001).

Map 5.1   Study area located near Jaipur in Rajasthan, India

Like much of Rajasthan, the people in this area are groundwater-dependent for their irrigation and domestic needs. There is no government water supply, save sporadically functioning village hand pumps. All village residents, therefore, rely on tubewells for water. Tubewells tap aquifers of varying depth, volume and recharge capacity. They were first introduced by the British in the late 1800s (Indian Irrigation Commission, 1903) but did not come to dominate groundwater supply until after they were promoted in 1960s green revolution development programs. Today there are at least 1.4 million, unregulated, agricultural tubewells in the state (Government of Rajasthan Groundwater Board, 2003).

In a number of respects, Rajasthan is an ideal place in which to study the process of environmental subject making. There is no history of formal state groundwater regulation, little historical state involvement in groundwater supply, and the resource is highly utilized, making it a facet of the environment that everyone has a particular prior understanding of, and something that people feel strongly about. This salience allows, furthermore, for the examination of differing viewpoints that are well grounded in practical experience. Following Agrawal, there are 'variations in the transformation' of subjects (2005: 186).

The problem of water scarcity

The world is facing a severe water crisis. According to the United Nations, groundwater use for irrigation by the world's farmers exceeds natural recharge rates by at least 160 billon cubic meters per year so that by 2025, 50 per cent of the world's population will face water scarcity (Rosegrant et al. 2002). In India, where groundwater meets 70 per cent of the country's irrigation needs and 80 per cent of its domestic water supplies, demand for both rural and urban uses is expected to exceed supply by 2020 (World Bank, 2005). This is of particular concern in the state of Rajasthan, where groundwater is an increasingly important source of both irrigation and domestic water. In 1990, 60 per cent of Rajasthani gross irrigated area was irrigated with groundwater (Directorate of Economics and Statistics, 1997). By 2000, nearly 71 per cent of irrigated area was groundwater-dependent. This includes an increase in net irrigated area between 1990 and 2000 of over 1.27 million hectares in the state (Directorate of Economics and Statistics, 2003). Further, 70 per cent of Rajasthan's population is groundwater-dependent for its drinking water (Black and Talbot, 2005). The alarm bell is ringing in the state as groundwater extraction currently surpasses recharge by 409.65 million cubic meters per year, resulting in falling groundwater levels in many areas by as much as 60 meters (Directorate of Economics and Statistics, 2003; Government of Rajasthan Groundwater Board, 2003). The groundwater situation is dire and there is a legitimate need for rethinking groundwater governance and development. It is not a question whether to regulate or not, but the particular character these reforms will take.

Rajasthan groundwater bill

The groundwater situation has sparked serious debate in Indian and Rajasthani development and government circles around the question of regulation. The Indian Constitution grants individual states the authority to regulate groundwater and currently Rajasthan has no groundwater regulation. And while the central government does not have the authority to regulate water resources, it has a history of providing guidance. For example, the Ministry of Water Resources drafted a model National Water Policy of 1992 and then updated it in 2002. More recently they drafted the 'Model Bill to Regulate and Control the Development and Management of Ground Water, 2005'. Both of these were intended to serve as a 'model' of surface water and groundwater regulation for states. All three of these bills encourage state governments to set up a 'Groundwater Authority' (hereafter Authority). Following the Central Government's 2005 'draft bill', the State of Rajasthan, working with multiple development agencies such as the World Bank and European Commission, and Indian NGOs (e.g. the Maharaja of Jodhpur's Jal Bhagirathi Foundation), formed an 'Expert Committee' to prepare the 'Rajasthan Groundwater Rational Use and Management Act of 2005' (the Act).

The framing of the groundwater problem and the provisions that the Act lays down follow from a recent World Bank report titled, 'India: Bracing for a Turbulent Water Future.' The report identifies two major problems exacerbating the groundwater problem: 1) 'indiscriminate pumping of groundwater' mostly for irrigation by farmers, and 2) 'provision of free power' in the agricultural sector (Briscoe 2005, for a short summary see World Bank, 2005). The World Bank's proposed solutions are uncomplicated in their presentation and are based on four market-based principles. The first is defining and setting water entitlements – transferable rights over water. Closely related to entitlements is the second principle of clearly defining property rights over water. The third is 'increasing supply and efficiency through technological expansion', including more efficient irrigation systems and surface water dams. And the fourth is establishing water user associations, thereby localizing governance. The World Bank is supporting its recommendations through an increase of rural water sector loans from $250 million between 1999 and 2004, to $1,400 million between 2005 and 2008 (Briscoe 2005).

The Act follows the World Bank recommendations, highlighting ownership, pricing, local governing bodies and technological expansion as the foundation for groundwater reform. But the Act also emphasizes setting up a 'State Ground Water Authority', a regulatory mechanism, and the dissemination of awareness and knowledge relating to water issues. The Expert Committee agrees with these provisions, but the subsequent Act is vague on a number of issues. These include determining entitlements (for example, would a water entitlement be based on land holdings or household size?), setting the price of water, and how to compose and form user groups. Under the Act, the Authority has the power to define and settle these points. The rest of this section focuses on the composition and powers of the Authority, especially its hierarchical structure and its mandate to disseminate knowledge and awareness of groundwater conservation. This is very similar to the forestry department efforts detailed by Agrawal in the effort to subjectify in terms of the environment.

The Act appoints a 'Chief Ground Water Officer' and forms a three-tiered hierarchical Ground Water Authority: the first tier is the 'Rajasthan Groundwater Authority' at the state level, composed of seven high-level elected officials and 15 appointed members; the second tier is the 'District Groundwater Authority', composed of two elected officials and ten appointed members; and the third tier is the 'Block Ground Water Authority', composed of one elected official and eight appointed members. These new Authorities are proposed not to replace the current bureaucracies charged with groundwater research and development but are in addition to them and draw on their personnel. For example, the Rajasthan Groundwater Authority is composed of the Chief Engineer for the Rajasthan Groundwater Board and the state's Chief Groundwater Officer. The Rajasthan Groundwater Board is responsible for mapping groundwater, collecting groundwater quality data and locating well sites. The Authority also includes one appointed representative from each the Human Rights and Women's Commissions, and several state secretaries: Agriculture, Industries, Irrigation, Power, and Forest. Therefore, the Authorities are disproportionately composed of appointed rather than elected members. Similar to forester skepticism in Agrawal (2005: 84) that villagers could not manage their own forests, this shows that the Expert Committee drafting the Act does not want to fully entrust groundwater management to local bodies, calling into question whether the goal is to decentralize and democratize governance, or to increase state control through decentralization.

Each tier of the Authority is allocated its own set of 'duties, functions, and powers'. The proposed Rajasthan Groundwater Authority would have 28 of these in addition to a mandatory meeting once every three months. Two of these powers, 'to supervise and control the District Ground Water Authority' and 'to spread knowledge and awareness regarding water issues', show the hierarchical top-down structure of decentralization as well as the goal to subjectify farmers with respect to environmental awareness (Expert Committee, 2005). This is further exemplified in the District and Block Groundwater Authority functions. The District Ground Water Authority has seven functions, two of which include: 'to follow and comply with the directions of the Rajasthan Ground Water Authority' and 'to spread knowledge and awareness regarding water issues'. And the Block Ground Water Authority has eight functions, three of which include: 'to carry out the directions of the District Ground Water Authority', 'to spread knowledge and awareness regarding water issues', and to advise the Gram Panchayats (locally elected decision making bodies) 'to properly utilize funds which are made available to them for the purpose of water conservation' (ibid). The Authorities, in addition to being predominately appointed rather than elected, function from top to bottom. The state level directs the district level and the district level directs the block or local level. The hierarchical structure and the mandate to carry out groundwater awareness activities to subjectify in terms of the environment, ensure that the demands of the state are carried out at the local level, but do not provide for a democratic process. The Forest Department's success in Agrawal's research, stemmed from the incorporation of peasant demands, albeit after much protesting. Recent state actions in groundwater indicate that rather than incorporating existing ways of knowing, practicing and institutionalizing groundwater use, the state plans to try to alter the existing, time-honoured, ways people think about groundwater, its use, and management to facilitate its regulatory efforts.

Making groundwater subjects

Farmer protests over inadequate irrigation water have become commonplace in Rajasthan, and the outcomes have often been violent. In October 2004, farmers in Sriganganagar District protested against not receiving their share of irrigation water from the Indira Gandhi Nahar Pariyojana project. Six were shot and killed by police. In June 2005, farmers in the Tonk District protested against not receiving their share of irrigation water from the Bisalpur dam project. Five were shot and killed by police. In February of 2005, I attended the conference 'Groundwater Management in Rajasthan: Issues, Perspectives and Policy' in Jaipur, Rajasthan, organized by UNDP and the Jal Bhagirathi Foundation. The Chief Minister of Rajasthan, Vasundhara Raje, gave the keynote address. With recent protests as the backdrop, she said in relation to future groundwater management:

> The government cannot do everything. It has to be a people's movement. I don't want to give scope for any anti-Government movement on this count. In the past we had gone out of the way in providing everything they asked for and we were taken by surprise, then I realized that it was absolutely important to be in touch with the people. The future is

going to be terrible unless this aspect is taken care of. The growing groundwater crisis in Rajasthan can be stalled only if every section of society supports the urgent task of building mass awareness about the need for conservation and regulation of groundwater resources (Raje address, 25 February, Jaipur GW Conference).

Much like peasant forest users through the early 20th century, therefore, Rajasthani farmers find themselves outside of the regulatory process and are protesting their lack of access to government and to water. Similar to provisions in the Act to 'disseminate knowledge and awareness', the Chief Minister also believes that people's perceptions must be changed. As she makes clear, the state is also trying to transform them into self-regulating subjects. As she went on to state: 'If people become conscious then we can even combat [a] situation like drought' (Raje, quoted by Times News Network, 2005). But statements such as these reveal something more ambivalent in government thinking. On the one hand, it implies that the current groundwater situation is the people's fault and if the state can just get people to become conscious and take responsibility then the problem can be solved. But on the other hand, the state does not integrate people into the regulatory decision making process. This is apparent both in the suppression of farmer protests and in the non-representative character of the institutions set up by the Act. Contrary to Agrawal, the state is not yet incorporating local input into the regulation (on which successful regulation depended) but as in Agrawal, the state is trying to condition peoples' thinking to be in line with state needs, and this may or may not benefit local people.

Even though the Act has yet to be adopted, in December 2005 Rajasthan's 'Expert Committee', who drafted the Act, orchestrated and launched the statewide *Jal Abhiyaan* (water awareness campaign), as a result of their desire to subjectify groundwater users towards conservation. Conservation outreach, such as the *Jal Abhiyaan*, is one of the duties of the Authority, but because the Authority and the Act have yet to be enacted, the Expert Committee carried forward this provision in an attempt to cultivate a sense of conservation ahead of the Act's formal adoption. At the center of *Jal Abhiyaan* was the *Jal Chetana Yatra* (water awareness march), a statewide march that began in May 2006 and went through 18,000 Rajasthani villages (*The Hindu*, 2006). *Jal Chetana Yatra* tried to raise awareness of groundwater use, protection, conservation and pollution, while informing farmers of more efficient irrigation practices. A second goal of *Jal Abhiyaan* was its statewide Rainwater Harvesting Campaign. Utilizing television, radio and newspaper advertising, as well as other outreach activities, it attempted to train villagers to construct 100,000 rainwater harvesting structures across the state by June 2006. The Campaign, furthermore, was not just about building rainwater harvesting structures; it was also about cultivating a sense of desire and need in groundwater users to do so.

Prior to this campaign, I spent a great deal of time in the field attempting to understand what Rajasthani farmers thought of the present water situation and government actions to combat it. I found many tensions. In the next section, I first compare the proposed institutional changes in groundwater governance with those in Agrawal's case. Then, I turn to the tension between government expertise and local knowledge and practices to demonstrate the high level of distrust that local people have for government expertise and authority. I illustrate these divergences by focusing on the variations between local and government tubewell location techniques.

The Mirage of Decentralized Groundwater Governance:
Rifts in Local and State Groundwater Expertise

It would be expensive, inefficient and politically costly to regulate groundwater without enlisting citizens in conservation efforts, which explains the government's attempts to subjectify them in terms of the environment. Similar to the forest peasants in Agrawal's study with respect to their access to forest resources, Rajasthani farmers are already protesting insufficient water supply. Further government crackdowns, such as the police shootings of 2004 and 2005, would be counterproductive and politically detrimental in a state where elections hinge on opinion in the agrarian sector (Corbridge and Harris 2000). Following the Forestry Department's lessons, Rajasthan's Expert Committee is attempting to regulate groundwater by creating governmentalized localities in the form of the Authority through subjecting groundwater users towards groundwater environmentalism. This could put farmers in a policy making and enforcing position, but the current structure of the Act does not ensure this. It reproduces the hierarchical structure of the state, while ensuring that existing channels of state power (e.g. the Groundwater Board's prominence) are maintained. Unlike in Agrawal where rule making is partially democratized, the Act institutionalizes a process to localize state power through its three-tiered Authority, each responsible for carrying out the dictates of the upper level. This would lead to the creation of 'governmentalized localities' where autonomy from the state would be 'imagined' indeed. Again, Agrawal designated 'imagined autonomy' for the process whereby villagers adopted the goals of regulation as their own, but which actually came from the state. Finally, critical to the Act, is its mandate to spread environmental awareness. The 18,000 village Water Awareness March, advertisements, and the state-wide Rainwater Harvesting campaign are major initiatives to subjectify (rural and urban) populations towards groundwater environmentalism, transforming them into self-governing subjects. But the effectiveness of this process will depend on the situatedness of individual as well as their historical interaction with the state.

The Rajasthan Groundwater Board is responsible for monitoring groundwater quality and quantity, and provides well location services to farmers. However, as groundwater becomes scarcer they are changing their focus to water conservation. They also sit on the Expert Committee responsible for drafting legislation. The transition to a policy-making role makes sense from the standpoint that they are familiar with the groundwater situation, but the Groundwater Board's traditional role is redundant. Groundwater is sufficiently mapped and there are few highly productive wells left to place. The future of groundwater bureaucracy in Rajasthan is in conservation and this is why the Groundwater Board is remaking itself in the image of conservation expert. But Rajasthani farmers historically intersected with the Groundwater Board in their role as groundwater-monitor and well-locator. These historical intersections have implications for the efficacy of the Groundwater Board's outreach activities, such as the 'Water Awareness March', Rainwater Harvesting Campaign and future governance. How do the Groundwater Board and farmers view each other based on their historical interactions? Examining these interactions is one lens through which to understand how these groups interacted, but also predict how they might interact in the future.

With these roles of the Groundwater Board in mind, rather than asking people what they thought of the Groundwater Board, I queried 151 farmers about whom they would seek advice from before digging a tubewell. Their consultancy options include: 1) government engineers from the Groundwater Board, who base their recommendations on historical hydro-geological maps and monitoring wells in the area; 2) tubewell drilling firms, who base their recommendations on accumulated experience in the area; and 3) *Sunghas* – local water diviners that utilize traditional Hindu astrological principles.

I was intrigued but not surprised by their responses. None indicated that they would consult a government engineer, 25 per cent indicated that they would consult a *Sungha*, 64 per cent indicated that they would consult a tubewell drilling firm, and 12 per cent indicated that they would consult both the *Sungha* and the tubewell drilling firm. Nobody would consult the government engineer, the future agents of groundwater conservation. The question is *why*? Why do farmers choose to rely on tubewell drilling firms and Hindu spiritual experts rather than government expertise? There is widespread trust in tubewell driller and *Sungha* abilities because their expertise is derived through practice, and by knowledge of and experience with the area, much like that of the farmer. Peasant farmers identify more easily with *Sunghas* and tubewell drillers than the government engineer, who relies on formal expertise (and is not afraid to exert it). Part of the answer lies in the following farmer quote:

> If we hire the government engineer [for tubewell placement], they don't know our area and we pay them rupees 2,000 and they don't come back. Also, they have all these formalities that we don't understand. The *Sungha* is more accurate and cheaper. The *Sungha* is only charging rupees 50 [about $1]. The *Sungha* is better than the engineer because the engineer doesn't care or know about the cost of drilling a tubewell, where the *Sungha* cares more and is more sympathetic. The engineer relies on physical methods that are not good for all areas. *Sungha's* methods are not only for this area, they are more universal and could be used anywhere; he could find water in Ajmer, for example (Farmer – 22 July, 2005).

This quote sheds light on four issues that farmers have with government engineers and the particular form of expertise they embody: 1) the government is unreliable; 2) the engineer is more expensive (there is the cost of the service and at least as much in bribes); 3) the engineer's methods are thought to be less accurate because their knowledge is regarded as particular, where the *Sunghas'* knowledge is thought to be universal and; 4) that the government does not really care about their plight. Therefore farmers do not believe in the engineers' methods and, while not represented in this quote, they do not believe in their results either – 80 per cent of survey respondents indicated that the *Sungha's* results were better than the engineers'.

This rift is further exemplified, but in the reverse, through the words of two engineers with the Groundwater Board:

> They [farmers] hire these *Sunghas* who walk around the field until they get the *feeling* and then they tell them to place the well there. There is no science in this. Their wells work sometimes, but they are not the most productive wells. *It's luck* (Groundwater Board Engineer – 20 May, 2005, speaker's emphasis).

> Farmers don't understand technical language (Groundwater Board Engineer – 2 August, 2005).

These engineers' sentiments question Rajasthani farmers' traditional methods and abilities to manage groundwater and their ability to understand the language of science (e.g. conservation). Much like the Forest Department skepticism towards forest users, the engineers' thoughts are in contrast to the official policy position that decentralized regulation is the solution. Groundwater Board engineers are on the 'Expert Committee' who drafted the Act, which accounts for its top-down hierarchical structure. Furthermore, mutual distrust between farmers and the Groundwater Board engineers will diminish state abilities to subjectify environmentally as it is the Groundwater Board engineers who are the first line of government in spreading environmental awareness. So too, it could result in further conflict over water resources. In sum the rift in environmental knowledge between farmers and the state, will result in the failure to institutionalize governmentalized localities and environmental subjectivities. The ultimate consequence will be, therefore, that future efforts to regulate groundwater in an equitable and more sustainable manner will be frustrated. But, as with the peasant protests in Agrawal, farmer protests could also result in revising the Act to make government more responsive to groundwater users (i.e. more democratic), while incorporating existing groundwater knowledges and institutions, making the policymaking process more inclusive.

Conclusion

This chapter applied Agrawal's environmentality framework to understand recent state efforts in Rajasthan, India to regulate groundwater through decentralized governance institutions. It found that the state is trying to produce individuals with environmental subjectivities in groundwater to further its groundwater regulation goals. Centralized regulation is too politically, socially, and economically costly. If regulation is going to be effective, the state understands it must foster decentralized governance institutions, such as environmental subjectivities, making groundwater users self-regulating and active participants in the regulatory process. It does this through various water awareness campaigns designed to instill a sense of groundwater scarcity and a need for conservation. But the state does not incorporate local groundwater knowledge and institutions into the proposed regulation. Instead it sets up a hierarchical regulatory mechanism that does not decentralize governance, but creates a means to govern localities from the center, while imposing its own sense of environmentalism. These proposed 'governmentalized localities' are similar to Agrawal's case in that they seek to create a sense of environmental awareness of and 'imagined autonomy' over regulating local resources but they diverge in that the state's vision does not incorporate groundwater users into a decision making role, thus retaining the power to regulate water with the state.

It exemplified the tensions in this process by looking at the divergences between local and state groundwater knowledge and perceptions in well location, which indicate both future resistance to government regulatory and conservation agents (i.e. Groundwater Board engineers). Finally, it concludes that these tensions will impede the future regulation of groundwater in Rajasthan, which is in a serious state of over exploitation. One possible way out of this downward spiral, is to first rewrite the existing Act to make it less hierarchical, which would make it more democratically responsive from the bottom-up. And second, there is a need to incorporate existing groundwater knowledge, practices and institutions to help foster a sense of conservation that may already exist. But if state repression of recent water user protests is any indication of future cooperation, this process has a long way to go indeed.

Acknowledgements

This project was funded by a Fulbright-Hays Doctoral Dissertation Research Abroad (DDRA) grant. The Graduate School at The Ohio State University supported the analysis and write-up of fieldwork findings through a Presidential Dissertation Writing Fellowship. I am grateful for the cooperation of many farmers, government agencies, and other people knowledgeable about groundwater and irrigation in Rajasthan, India. I owe a particular debt to my research assistant Jaywant Mehta. I would also like to express gratitude to the School of Desert Sciences in Jodhpur, Rajasthan for their support. Finally, I gratefully acknowledge the intellectual support provided by Kevin Cox and Paul Robbins. Any shortcomings are my own.

References

Agrawal, A. (2005), *Environmentality: Technologies of Government and the Making of Subjects* (Durham and London: Duke University Press).

Black, M. et al. (2005), *Water – A Matter of Life and Health: Water Supply and Sanitation in Village India* (New Delhi: Oxford University Press).

Briscoe, J. (2005), 'India's Water Economy: Bracing for a Turbulent Future' (New York, World Bank).

Corbridge, S. et al. (2000), *Reinventing India: Liberalization, Hindu Nationalism and Popular Democracy* (Malden, MA: Blackwell).

—— (2005), *Seeing The State: Governance and Governability in India* (Cambridge: Cambridge University Press).

Directorate of Economics and Statistics (1997), 'Statistical Abstract – Rajasthan 1995' (Jaipur, Directorate of Economics and Statistics).

—— (2003), 'Statistical Abstract – Rajasthan 2001' (Jaipur, Directorate of Economics and Statistics).

Expert Committee (2005), 'Rajasthan Groundwater (Rational Use and Management) Act, 2005', Rajasthan High Court Bar Association.

Foucault, M. (1991), 'Governmentality', in Graham Burchell, et al. (eds), *The Foucault Effect: Studies in Governmentality* (Chicago: University of Chicago Press).

Government of Rajasthan Groundwater Board (2003), 'Personal Communication', in Birkenholtz, T. (ed.), *Jaipur*.

Indian Irrigation Commission (1903), *Report of the Indian Irrigation Commission, 1901–1903. Part 1 – General* (Calcutta, Office of the Superintendent of Government Printing, India).

Jeffrey, C. (2001), '"A Fist is Stronger than Five Fingers": Caste and Dominance in Rural North India', *Transactions of the Institute of British Geographers* 26, 217–236.

Martin, A. (2003), 'On Knowing What Trees to Plant: Local and Expert Perspectives in the Western Ghats of Karnataka', *Geoforum* 34, 57–69.

Morris, C. (2006), 'Negotiating the Boundary Between State-led and Farmer Approaches to Knowing Nature: An Analysis of UK Agri-environment Schemes', *Geoforum* 37, 113–127.

Robbins, P. (2000), 'The Practical Politics of Knowing: State Environmental Knowledge and Local Political Economy', *Economic Geography* 76, 126–144.

Rose, N.S. (1999), *Powers of Freedom: Reframing Political Thought* (Cambridge: Cambridge University Press).

Rosegrant, M.W. et al. (2002), 'Global Water Outlook to 2025: Averting an Impending Crisis', Washington, United Nations – 2002 Johannesburg World Summit on Sustainable Development.

Scoones, I. et al. (1994), 'Knowledge, Power and Agriculture – Towards a Theoretical Understanding', in Scoones, I. et al. (eds), *Beyond Farmer First: Rural People's Knowledge, Agricultural Research and Extension Practice* (London: Intermediate Technology), 16–32.

The Hindu (2006), 'Govt.'s "Jal Chetana Yatra" to Begin from May 15', *The Hindu*.

Times of India (2005), 'Raje Discusses Steps to Enhance Groundwater', *Times of India*.

World Bank, (2005), 'India's Water Economy: Bracing for a Turbulent Future', http://www.worldbank.org.in/WBSITE/EXTERNAL/COUNTRIES/SOUTHASIA EXT/INDIAEXTN/,,contentMDK:20668501~pagePK:141137~piPK:141127~theSitePK:295584,00.html (home page), accessed 3 January 2005.

Chapter 6

Discursive Spearpoints: Contentious Interventions in Amazonian Indigenous Environments

Logan A. Hennessy

People see what they want to see in native cultures. They can see vicious barbarism and pagan idolatry, or they can see wise, understanding, noble savages living in harmony with animals and plants as well as other humans. Our vision of native peoples ... often reveals more about ourselves than it does about them (Weatherford 1994).

Introduction

The Huaorani are a small population of 1,800 indigenous people in Ecuador's Amazon forest. The recent Hollywood film, *End of the Spear* (Gavigan et al. 2006), is based on a true sequence of events that led to contact and missionary pacification of the Huaorani in the late 1950s. For the most part, the dramatization of Huaorani violence and subsistence practices in the film can be substantiated by ethnographic reports (Rival 1992, 1996; Robarchek and Robarchek 1998; Lu 1999; Yost 1981b; Yost and Kelly 1983). The sequence of missionary activities, first episodes of contact with the Huaorani, and the actual 'cast of characters' in the film are also consistent with first-person stories (Kingsland 1980; Wallis 1960).

What is difficult to ignore when watching *End of the Spear*, however, is the film's presentation of its setting as if the movie was occurring in a spatial and historical vacuum. Ecuador's eastern rain forest region, which is called the *Oriente*, is one of the most contested environments in the entire Amazon basin. Indigenous resistance to various forms of development in this place have catalyzed a complete transformation in national politics marked by oil rig takeovers in the forests, multi-million dollar lawsuits against foreign oil corporations, widespread protests, and militant coups in the capital of Quito. Many of the struggles over resources and environmental and cultural protection have played out in Huaorani territory. In fact, they have been one of the most formidable groups to reckon with among all of Ecuador's indigenous peoples, yet in no point of *End of the Spear* is any of this history of resource conflict given due coverage.

What is clear is that the film produces a reductionist discourse of indigenous savagery to illustrate the need for a particular sort of intervention in Huaorani territory. This is evident from the opening scene which powerfully establishes the brutality of Huaorani inter-tribal warfare by reproducing a spear-raid. From the darkness of the rain forest several warriors emerge to surround another Huaorani house and greet the unsuspecting family with a shower of spears and machetes. This and other scenes from the film depict the Huaorani as ruthless murderers whose internal conflicts desperately need divine intervention.

It is also true, however, that part of their reputation for fierceness emerged from finding not just other Huaorani or missionaries, but also Andean settlers, the neighbouring Quichuan peoples, and oil surveyors at the ends of Huaorani spears. In this gap between essentializing Huaorani violence as a cultural anomaly versus situating it in the political economic context of resource and territorial struggles can be found the roots of appropriating Huaorani identity for specific development purposes. Thus, the ways in which a number of current groups – conservation NGOs, the State, and oil companies – have discursively constructed Huaorani identity underwrites the actions and consequences of their physical interventions into Huaorani space. Here, then, the question is how these multiple and powerful 'outside' groups essentialize the Huaorani in different ways, and what the emergent consequences of these competing simplifications might be on the Huaorani's environmental politics and mobilization.

In this chapter, I argue that the material struggles for space and resources in Huaorani territory create ideological, competing productions of essentialized identities that paradoxically exacerbate ethnic differentiation. Building on earlier fieldwork in Huaorani communities and Ecuador's *Oriente*, I develop a short history of contact and regional settlement patterns to illustrate variations in Huaorani cultural-ecological practices that existed prior to the 1990s. This is necessary to provide a background for where the Huaorani were spatially and culturally prior to a decade of particularly intense conflicts in which subsequent discursive interventions are analyzed. The focus will then shift into comparative discourse analysis of various groups intervening in Huaorani territory during the 1990s. I review the profiles of four tactics employed by non-Huaorani actors over this period to direct the Huaorani's social and environmental development. I categorize these productions in terms of the 'fixing' of the Huaorani's spatial-cultural diversity as abstractions of time, space, or nature. These profiles also construct narratives of either socio-ecological decline or progress that substantiate political leverage for managing the land and resources in Huaorani territory to the distinct benefit of these outside actors and institutions.

The approach in this research is very much informed by the interface between indigenous identity theory, regional discursive frameworks (Peet and Watts 2004), and the debates over non-material constructions of place and space. The next section engages some of this literature with a view toward recognizing the discursive practices of state and global forces in constructing 'strategies of global localization' (Escobar 2001: 161) in indigenous environments. In this framing, I am expanding the analytics of Escobar's (2001: 161) sense of how 'capital, state and technoscience engage in a politics of scale that attempts to negotiate the production of locality in their own favor,' by interrogating multiple constructions of a specific indigenous group inhabiting a particular place. Instead of accomplishing uniform industrial, environmental, or indigenous rights goals, however, my research shows that these 'discursive spearpoints' ironically undermine indigenous ethnogenesis by perpetuating internal divisions. This is where the Huaorani case makes an important contribution to the ongoing debates over the defence of place and 'areal differentiation' (Castree 2004: 138) by local and disempowered (indigenous) groups. This chapter provides evidence for how evolving efforts of indigenous exclusion are compromised by the multiplicity of political-environmental factions among intervening groups.

Identity Studies in Indigenous Social Movements

Much of the existing research on indigenous environmental politics is concerned with the source and authenticity of indigenous peoples in roles as environmental stewards, defenders of static nature, or passive victims of development (Brosius 1997a, 1997b; Conklin and Graham 1995; Li 2000; Ramos 1992; Slater 1995; Ulloa 2005). Out of these challenges to indigenous environmentalism, a key debate emerged over the production and employment of identities as political tools in the struggle for resources. While recent studies of indigenous identity 'articulation' (Li 2000), particularly in Ecuador (Bebbington 2004; Perreault 2003; Valdivia 2005), have importantly documented community narratives and discourses, this chapter is more concerned with the nature in which multiple 'external' groups develop competing discourses of indigeneity to control indigenous environments. In this context, it is my claim that the practice of environmental politics among non-indigenous actors still remains undertheorized.

The early work on indigenous discourses tended to favor the political power to produce indigenous identities through the imagination of foreigner's stereotypes (see Slater 1995). Conklin and Graham (1995: 702) support this point when they say 'Indianness and signs of Indianness have a symbolic value that is not intrinsic but bestowed from the outside'. Alcida Ramos (1992: 14) similarly argues that indigenous identities are 'hyperreal' because they are idealized abstractions produced by environmental NGOs as 'a model that molds the Indians' interests to the organization's shape and needs.' These works clearly argue that identities primarily emerge from external constructions rather than from productions of the indigenous group themselves. More recent research, in Ecuador in fact, has further shown how global players simplify problems of ethnodevelopment through generic understandings of indigenous places (Laurie et al. 2004: 477) that fail to represent the evolving and occasionally contradictory realities of local and indigenous environmental politics (Perreault 2003: 598).

Complementing earlier studies, this chapter offers a more explicit schematic of *how* oil companies, missionaries, environmentalists, and the state use discourse and symbolism to intervene in the production of specific indigenous spaces in Ecuador. In this realm, Sawyer's (1997, 2004) detailed account of the ARCO oil company's manipulation of Quichuan organizations to the south of Huaorani territory is key; her studies illuminated the edges of a larger canvas on which multiple profiles of indigenous inhabitants are created and 'sold' by external actors. Consequently, there is an opportunity to understand indigenous environmental politics as a field of practice among several players simultaneously contending for indigenous space and resources. It is furthermore important to draw connections between the confluence of these competing discourses and their emergent consequences for indigenous peoples and their livelihoods.

In the subsequent sections I present a case study of the Huaorani people focused on various outside discursive interventions in conflicts over oil, environmental protection, and indigenous rights in the 1990s. This framing questions how non-Huaorani groups use politics, language, and the media to construct abstracted notions of Huaorani indigeneity and territory to the benefit of these outside and powerful actors. In the process, I expose an underlying pattern of diverse cultural adaptations to these constructions and the internal tension of the Huaorani people created by multiple appropriations of their identity.

Early Constructions and Transformations of Huaorani Space

At the point of first contact in the 1950s, the Huaorani population was estimated to be between 500–700 members spread out over 21,000 km^2 in Ecuador's eastern lowlands (see Figure 6.1) (Rival 1994; Robarchek and Robarchek 1997; Yost 1981a, 1981b). Prior to this point, the Huaorani epitomized what Castree (2004: 136) calls 'exclusionary localism' – the pursuit of a differential geography 'to make their own places, rather than have them made for them.' They repelled any infringements into their territory through violent isolationism. Although immediate Huaorani neighbours were usually friendly, internal warfare with more distant households and with non-Huaorani indigenous groups was frequent (Rival 1996, 2002; Robarchek and Robarchek 1997; Yost, 1981a, 1981b). From an ethnographic standpoint, this warfare complex was partially comprised of superstitions of witchcraft, conflicts over arranged marriages, and a fixation on revenge (Robarchek and Robarchek 1997).

A complementary explanation for the Huaorani's violent isolationism rests in the hostilities of Ecuador's Amazon region. Huaorani interactions with outsiders prior to the 1950s included encounters with rubber tappers, missionaries, colonists, neighbouring Quichua and Shuar peoples, and geological surveyors for the Shell Oil Company (Robarchek and Robarchek 1997: 20–24). Ostensibly none of these meetings were peaceful, but fell into one of two types of clashes: either they were coincidental meetings in the forest that resulted in immediate attacks, or else they were calculated, deliberate raids from either side. On one occasion, Huaorani were captured and enslaved on haciendas. In another, twelve employees of Shell's exploration team are thought to have been killed by the Huaorani in 1937 (Robarchek and Robarchek 1997: 23–24).

Through these episodes of violent encounters with outsiders, the Huaorani became renowned as some of the most feared indigenous peoples of the entire Amazon. The settlers, officials, and travellers of the region called them the *Auca*, which is a Quechuan word for 'savage ... violent, magical, and monstrous people' (Taussig 1986: 97). This term spread among Andean society for centuries as a racial slur to describe indigenous people from the Amazonian lowlands (Taussig 1986: 97). In Ecuador, however, this name was given to the Huaorani as a mark of distinction. *Auca* was their official ethnic classification up until the late 1970s. It specifically ascribed a negative construction of indigeneity to a particular group of people inhabiting a particular place.

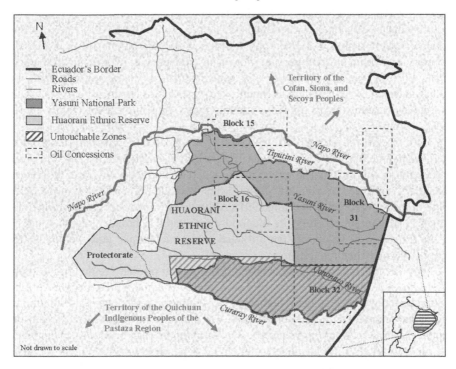

**Figure 6.1   Map of Huaorani territory and surroundings, Ecuador
(adapted from Kane 1995 and Proyecto PETRAMAZ 1999)**

The Huaorani's reputation as a ruthlessly savage people was entirely a foreign construct circulated by state officials, explorers, farmers, and the Huaorani's immediate indigenous neighbours. The *Auca* construct additionally established fixed indigeneity to the landscape, in which rivers signified the hard boundaries of a savage place marked by violence. South of the Napo river was '*Auca* territory', home of '*los infieles*, the heathens,' land of murderous savages (Blomberg 1957: 13). In this sense, both the Huaorani and their territory embodied Ramos' concept of the 'hyper-real Indian'.

Through the help of Dayuma, a converted Huaorani woman who fled her group, eventually peaceful contact was accomplished by a group of evangelical missionaries in 1958. This story of pacification and modernization that is glorified in *End of the Spear* (Gavigan et al. 2006). Shortly after peaceful contact, the Ecuadorian government awarded the Huaorani a Protectorate zone (Yost 1981b), but this comprised only a small fraction of their original territory. Throughout the 1960s, government and missionary leaders tried to attract remote Huaorani villages and lure them to the Protectorate in an effort to relocate the entire population and facilitate programs in religion, health, and education.

The modernist project of life in the Protectorate transformed the hunter-gatherer lifestyle of the Huaorani's nomadic isolationism dramatically. The Protectorate concentrated a formerly dispersed and independent population onto a smaller area resulting in the collision of formerly isolated family groups (Rival 1997, 1998). A few villages became centres of higher populations, new social relations, the development of more substantial agriculture, and increasing contact with outsiders and markets (Yost 1981b). The new authority of teachers and missionaries discouraged the hunter-gatherer lifestyle which eventually led to some loss of forest survival skills and ecological knowledge for those communities living in the Protectorate (Rival 1992, 1997).

The processes of internal differentiation of the Huaorani continued to unfold as the *Oriente* transformed from a primarily agricultural-based economy to one of oil production in the 1970s. The government signed a contract with a locally created subsidiary of Texaco oil company for 20 years of crude oil production in the fields north of the Napo River (Kimerling 1991). Although Texaco concentrated its activity outside of Huaorani territory, its monopoly facilitated a broader regional survey of the *Oriente's* oil fields through affiliated exploration firms, and eventually this activity reached Huaorani territory.

After Texaco financed a road connecting the central-northern region of Huaorani Territory with the frontier town of Coca, many Huaorani men began working on exploration crews in their homeland. This changed the Huaorani's cultural-ecological practices in a number of ways. For example, this new contact with the 'outside' world economy introduced a semi-consistent cash economy, encouraged the use of markets, and developed relations of dependency on foreigners for basic needs. Access to shotguns and ammunition allowed an increase in the frequency and intensity of local hunting over the traditional weapons of blowguns and spears (Yost 1981b, 1983). In this way, a number of large game species were brought to near extinction in the areas surrounding the larger Protectorate villages (Lu 1999). This created an increasing demand on external food sources (Yost 1981b), which created the need for more cash to buy food and shot (Sierra et al. 1999). The intense hunting yields, higher Huaorani populations, and sedentary households also gave rise to social tensions over local hunting grounds. The Huaorani adapted to this variety of cultural and economic influences by redistributing themselves across the landscape. Eventually some household groups moved east away from the Protectorate to return to a more traditional lifestyle of hunting and gathering (Kane 1995; Lu 1999; Mena et al. 2000: 76).

These socio-cultural transformations – modernization in the Protectorate, oil exploration and a cash-based economy, and the relocation of family groups – generated a gradient of Huaorani culture that differentiates the eastern remote villages from their western communities (Rival 1992, 2002; Robarchek and Robarchek 1997; Lu 1999). Certainly this is the case with the still isolated, uncontacted band of Huaorani households known as the Tagaeri and Taromenane. They are rumoured to be living in the southeast parts of Huaorani territory along the Curaray river, and to date, no outsiders have been able to make peaceful contact with them (Smith 1993: 89–121; Reyes and Villavicencio 1999). These decisions for participation in various forms of 'development' have created a range of cultural adaptations among the Huaorani people, from the modern villages in the west to the peacefully contacted traditional communities in the east, to the uncontacted, potentially dangerous bands located somewhere in the south.

By the late 1980s, this spatial and cultural heterogeneity of the Huaorani continued to evolve but the political climate over oil, the environment, and indigenous rights in lowland Ecuador increasingly polarized the identities of indigenous peoples. The environmental and indigenous rights movements in Ecuador and the United States were embroiled in the conflict over Texaco's history of contamination. The timing of this mobilization just prior to a proposal for drilling in the Huaorani region cannot be overemphasized. When the Ecuadorian State Petroleum Company, CEPE (Corporación Estatal Petrolera Ecuatoriana) introduced three contracts to develop oil fields in the Huaorani area in 1986, a new era of conflict began. Indigenous and environmental social movements now provided the structural capacity for organized resistance.

Their reputation as the fiercest tribe in the Amazon transformed the Huaorani into a *cause célèbre* in which much larger struggles over neoliberal policy, extractive industries, indigenous rights, and the protection of rain forests unfolded (Kimerling 1991; Watts 1999; Sawyer 2004). The contentions over Huaorani territory by intervening groups increasingly turned to constructing abstracted discourses of their identity. The conflict encouraged external groups to make broad but simplistic claims of who the Huaorani really were, and to use these identities to promote their own institutional agendas. Underneath the various filaments of identity cast by competing actors, the Huaorani's spatial-cultural differentiation continued.

Indigenous Environmental Contentions in Huaorani Territory in the 1990s

With the above historical mosaic of cultural landscape in mind, in the 1990s, a variety of media were used to represent who the Huaorani were, what kinds of relationships they have to nature, and how they fit into modern society. These constructions of identity were also deliberate political positionings in contentions over the control of land, resources, and Huaorani ethnodevelopment. The following profiles explore the ways these groups – again, NGOs, the State, and oil firms – constructed Huaorani identities for their own purposes. Taken together, they illustrate various appropriations of indigenous identities with two distinct and emergent properties.

First, the political terrain of conflict in which several groups are intervening for specific economic, social, or ecological interests – in some cases with diametrically opposed politics – creates a polarizing effect on Huaorani identities. Intervention hinges on constructing narratives of decline or progress, but rarely the status quo. In this sense, the Huaorani, the area's biodiversity, the notion of indigenous rights, or the economic value of fossil fuels – any variables of 'place' – all become conceptual fields for developing political platforms to achieve material access or control over Huaorani territory. These are actively produced and reproduced with positive or negative narratives in order to legitimate involvement. Each of these 'discursive spearpoints' makes either explicit or implicit statements about the Huaorani and the place they live. I provide four such examples illustrating various ways in which intervening groups 'fix' the Huaorani in relation to spatial, temporal, natural, and cultural variables.

Second, a by-product of producing contentious standpoints is that the Huaorani's variety of cultural adaptations are homogenized into single, coherent identities by intervening actors. There is a clear practice of ignoring 'the dynamic processes of indigenous self-identification and contestations that take place within [indigenous] communities', (Valdivia 2005: 295) in the images and discourse of intervening groups. Profound rifts in cultural, ecological, and political practices can be found among the Huaorani. This diversity is overlooked in external constructions of their identity. Despite these similarities, the research will show that intervening groups are actually simplifying Huaorani identities in very different ways.

Fixing indigeneity in space

An 'Action Alert' by the Rainforest Action Network (RAN) in 1997 discusses the isolated Huaorani clan, the Tagaeri, as threatened by Occidental Petroleum Corporation's development project in concession Block 15. The uncontacted Huaorani clan, the Tagaeri, are showcased in the alert as isolated Indians living in harmony with nature and roaming the regions *north* of the Napo River in desperate need of protection of their fragile, 'ancient' and 'sustainable' culture. There is little doubt that the main motive for creating this Alert was to organize against oil development; however, a close inspection of the Alert reveals one major problem. While not impossible, it is highly unlikely that the Tagaeri appeared north of the Napo; both Huaorani and government sightings have always located them much farther south, across hundreds of kilometers of forest, to the southeast from the end of the *Via Auca* (see Smith 1993: 89–121; Proyecto Petramaz 1999; Reyes and Villavicencio 1999; Aguirre 2006; El Comercio 2006a, 2006b; Hoy 2006). It seems that RAN utilized the mere *idea* of the Tagaeri clan as symbolic capital for promoting their environmental cause in an area not usually associated with Huaorani territory.

This action alert was nonetheless an effective campaign that invoked a declensionist discourse to thwart the construction of oil wells, even in a region quite removed from the known home range of the Tagaeri clan. This demonstrates the willingness of intervening parties to stretch Huaorani territory beyond its historic range in order to promote an institutional agenda of environmental conservation over a broad space. The effect is to fix Huaorani indigeneity to a larger spatial area overlapping both banks of the Napo river, and ascribing an identity of ancient culture to this place. With these discursive strategies, RAN is reducing the variability of Huaorani culture into a denominator of the Huaorani's uncontacted clans to politically communicate socio-ecological vulnerability.

Fixing indigeneity in time

The elasticity of invoking the discourse of harmony and sustainability in indigenous development is made clear by contrasting RAN's Action Alert with images used by the primary oil company operating in Huaorani territory in the 1990s, Yacimientos Petroliferos Fiscales (YPF). The similarities in discourse between the two are striking; like RAN, the YPF company utilized a similar image of indigenous harmony with nature.

In two of YPF's newspaper advertisements from 1999, a Huaorani child jumps through the forest under a headline that reads, 'a pact between man and the earth.' In both images, Huaorani territory is depicted as a manageable landscape capable of harmony between nature, industry, and indigenous peoples, with YPF as their heroic saviour. In one ad, the sub-caption reads *Hacer felices a los niños* (to make the children happy/to bring the children happiness), and on the other it reads *Proteger la Naturaleza* (to protect nature) (see Figure 6.2). These images reflect the tactics used by oil companies to project "illusions of betterment" in Ecuador's Amazon basin (Sawyer 2004: 8). The indication is that wise direction by the YPF company will yield a sense of harmony and unity between oil development, nature, and indigenous groups.

Figure 6.2   YPF newspaper advertisement (El Comercio 1999)

The explicit and implicit discourse of these ads reveal a rich cache of meanings that position the company in current debates about industrial development in indigenous occupied rain forests. Particularly in the second caption, *Proteger la naturaleza*, the public is led to believe that YPF is focused on protecting a temporally fixed nature and the Huaorani happen to be part of it. The perceived loss that would result without the intervention of the company is not a specific place (as in the case of RAN's approach), but a historical geography of a fragile ecological region that has the Huaorani people in it. The company's strategy is therefore to fix indigeneity in time.

Fixing indigeneity against nature

Another image and narrative of the Huaorani illustrates how they can be considered threats to nature and biodiversity. The Sustainable Uses for Biological Resources (SUBIR) project is an integrated conservation and development program with a variety of international financial support. SUBIR is working in multiple areas of Ecuador with the goal of demonstrating 'that long term biological conservation can be combined effectively with self-sustaining community development', (SUBIR 1999).

SUBIR shares an institutional belief with the government that the Huaorani's own management of natural resources is in need of assistance. In the words of one government official, 'We're here to help [the Huaorani] protect their own park because this park belongs to them' (Walker 1996). This belief rests on the view that the Huaorani should develop a biocentric ethic of conservation which can be encouraged through the intervention of wildlife ecologists. The SUBIR project was enlisted for this purpose, and their staff scientists led studies on the impacts of Huaorani hunting practices on local wildlife (Mena et al. 2000). Researchers concluded that the 'prime taxa [were] to become locally extinct', since there was 'no reason to believe that [the Huaorani] will change their current hunting patterns', (Mena, et al. 2000: 76–77).

This discourse of biological crisis slices off a reductionist view of Huaorani cultural ecology and uses it to construct a declensionist narrative of place. The differences between a number of Huaorani communities practicing a variety of survival and livelihood strategies are ignored. The resonant discourse in the SUBIR studies constructs a dichotomy between indigenous peoples and nature in which the Huaorani are seen as threats to surrounding biodiversity.

Whereas in the examples discussed earlier (RAN, YPF) different dimensions of indigenous territory are used as political capital for promoting institutional agendas, this example of intervention completely ignores narratives that draw deep connections between the Huaorani and and the Napo region of Ecuador. In fact, it has the opposite result of alienating the Huaorani from their own homeland. The SUBIR project neither fixes Huaorani indigeneity in relation to their temporal presence in the area (like YPF) nor ascribes a spatial aggregation of people and space (like RAN). Instead, the SUBIR ethic re-*places* Huaorani identity as something damaging to a natural world that pre-dates them instead of deeply embedding them in its primal roots. This grafting of Huaorani territory onto global models of conservation generalizes the highly complex and rooted local geography of place in terms of quantitative ecological measurements and statistical analyses.

Fixing indigeneity in exclusion

In February of 1999, Ecuador's President Jamil Mahuad Witt signed a new protected area decree creating two *Zonas Intangibles* (see Figure 6.1). Those 'untouchable zones' in Ecuador's Amazon basin restrict all extractive activity within their borders, including hunting, fishing, logging, and oil extraction (Proyecto Petramaz 1999). One of these areas is the southern stretch of Yasuni National Park and Huaorani Territory. The motives behind this action were 'to prevent further irreparable damage to the indigenous populations and the environment, thereby guaranteeing the protection of life in all of its forms' (Proyecto Petramaz 1999). But perhaps the most significant motive for creating this untouchable area is that this is where the uncontacted Huaorani clans, the Tagaeri and Taromenane, currently live.

The untouchable area was created by PETRAMAZ, The Project of Oil Exploitation, Environmental Management, and Sustainable Development in the Ecuadorian Amazon. A series of events and interventions over the last 40 years were attributed as the basis for the decree, indicating it was the resolution of several groups who had engaged in contentions over Huaorani space. While SUBIR promotes a rational rainforest management plan that problematizes the indigenous presence of the Huaorani, PETRAMAZ's *Zonas Intangibles* protected area promotes a sense of rational management through the preservation of an 'authentic' indigenous landscape. All of the intervening parties (except, perhaps, for the evangelical missionaries) felt as if protection of 'pure' indigeneity is consistent with their organizational agenda.

The PETRAMAZ project illustrates the transformation in intervention tactics, approaches to managing indigenous space, and productions of indigeneity over the last half century. Nearly 50 years after successful contact with most of the Huaorani, there are still uncontacted indigenous people living in the forests of Ecuador. The Tagaeri and Taromenane clans still possess a propensity for violent interactions with all outsiders, just like the Huaorani once did in the period depicted by *End of the Spear*. The approach to intervention with untouchable zones, however, is the opposite of contact.

Whither the Huaorani?

Comparing the discourse and symbolism utilized by outsiders reveals a range of strategies in the production of indigenous space and a certain elasticity of perceiving the Huaorani. The ways in which the Huaorani and nature have been socially constructed through time corresponds not only to popular conceptions of indigenous peoples, but also to specific actors with particular and conflicting political agendas. As much as these profiles demonstrate the conceptual basis for various territorial and ethnic interventions, they have largely been unsuccessful in a material sense. None of these interventions are complete – not the pacification, not the industrial development, not the conservation, not the territorial protection, and not the untouchable zone. Instead, each of these interventions has contributed to partial transformations of Huaorani space. Moreover, they reflect only a sample of the larger practice of intervening groups vying for control of the Huaorani and their territory. This variety of intervention not only reveals the ways in which simplified notions of Huaorani identity are groundless because they betray the complexity of Huaorani environmental politics, but it also suggests that collectively such contentions can be counterproductive to an already torn population.

The Huaorani are emerging into a political process of ethnogenesis from a historical context of dispersed households without structures of central authority (Rival 1992; Robarchek and Robarchek 1997; Lu 1999). Communities and villages are grouped throughout the territory in a spectrum of acculturation from west to east. These households have different access to education, markets, employment, forest resources, tourists, oil companies, missionaries, and health care. The process of ethnogenesis is complicated enough by these internal differences in cultural adaptations, but is further constrained by the additional imposition of competing identities by these outside actors.

The evidence for how contentions have exacerbated internal divisions can be seen in recent events since the 1990s. Much of the internal crisis is a result of unilateral agreements for new resource developments between a few Huaorani individuals and intervening groups. This is happening on multiple scales, and the consequences range from complications in tribal-level political governance to an escalation of violence in more localized disputes.

On two occasions, foreign firms successfully signed contracts with Huaorani leaders that were not approved by the larger population. The first dispute involved oil. In March 2004, the Huaorani signed an agreement with a Brazilian company, Petrobras, for permission to develop Block 31, but this administration was replaced and accused of acting in self-interest the following year (*El Comercio* 2005, 2006b, 2006c). The new Huaorani leadership denounced the accord as invalid and staged a protest at the company's Quito office in July of 2005. Eventually Petrobras abandoned its claim in 2006 (*Servindi* 2006). The second case involved biological resources. A US company, EcoGenesis Development Company LLC, persuaded a Huaorani leader to sign a usufruct agreement in 2005 for developing pharmaceuticals and forest products (*El Comercio* 2006b). The contract allowed the company 30 years of access to all the biological resources in Huaorani territory, but this was rejected by the larger Huaorani population who again removed their official leaders in a special election. Disagreements over events like these have caused the leadership of the Huaorani to change hands four times in the past six years.

Meanwhile, violence continues to plague the still uncontacted Tagaeri and Taromenane clans living inside the untouchable zone. Some of the contacted Huaorani communities living near the Tagaeri and Taromenane are allowing loggers to enter the zone (Aguirre 2006; Lucas 2006), but using their traditional spears, the isolated clans have repeatedly attacked and killed these intruders. Several violent exchanges have unfolded, most notably a Huaorani-led massacre of 20–30 of the Tagaeri/ Taromenane clan in June of 2003 (Aguirre 2006; *El Comercio* 2006d). Additional attacks with fatalities on both sides unfolded again in April of 2006 (Aguirre 2006; *El Comercio* 2006b, 2006c, 2006d). These episodes of internal fighting exemplify one of the more critical ways in which regional pressures for natural resources continue to fuel inter-tribal war on local scales. They definitely contextualize perennial sources of resource conflicts and violence that are inexplicably missing from *End of the Spear*.

These events are but a few examples of how external competition for who and what the Huaorani are has exacerbated tensions within the Huaorani population and diffused their ethnic identity and cohesion. The net effect of intervention is therefore divisive at a time when the Huaorani's struggles for cultural integrity and territorial defence require some degree of cohesion. It seems clear that the Huaorani's contribution to geographical difference (Castree 2004: 138) has been compromised by the discourse and actions of national and international players.

Conclusion

These findings reveal a case where emergent consequences of intervention undermine long-term goals of ethnic mobilization and the capacities of indigenous peoples to maintain distinct identities that are tied to particular places and, perhaps, their defence (cf. Escobar 2001; Castree 2004). The trouble starts when interventions poorly reflect the transformative links between identity, cultural differentiation, and territory, all of which are constantly being produced and reproduced by indigenous peoples themselves (Li 2004: 343; Castree 2004: 155). To 'fix' an identity as an outsider is to stabilize a moving target, to create temporal, spatial, or cultural boundaries around a people and a place that constrain evolving cultural adaptations. This point is all too clear for the Huaorani who have come to recognize the threats of interventions on their own. In a December 2005 Assembly, the newly elected Huaorani leaders equally criticized the presence and activities of NGOs and oil companies while insisting on political unity and conservation of their territory (El Comercio 2006a). This, it seems, has always been their own spearpoint. If Weatherford (1994) is correct in saying 'Our vision of native peoples ... often reveals more about ourselves than it does about them,' then avoiding further negative impacts on indigenous peoples must begin with views of people and places that acknowledge internal variations and the right to adaptive cultural ecologies of difference. For the Huaorani, this means an opportunity to make their own place.

Acknowledgements

This chapter was developed from a study of the history of intervention in Huaorani Territory. It is based on personal experiences travelling to Yasuni National Park and Huaorani territory in both 1995 and 1999. Kelly Swing at the Universidad San Francisco de Quito and the Tiputini Biodiversity Station, first brought me to the Ecuadorian lowland rain forests. His experience working with the Huaorani was inspiring and integral to my work. I would especially like to thank the Huaorani people for the opportunity to enter their territory, visit their communities, and listen to their stories. Professors Claudia J. Carr, Jeff Romm, and Carolyn Merchant, and colleagues Arielle Levine, Karen Levy, Elizabeth Havice, and Josh Dimon at the University of California, Berkeley all provided valuable feedback on the initial ideas and drafts of the material in this chapter. Finally, the editors deserve considerable praise for their careful scrutiny of initial drafts. Any errors are my own.

References

Aguirre, M. (2006), 'Ecuador: Hidden Indigenous Communities Fight Extinction with Spears', Inter-Press Service News Agency (published on-line 13 July 2006) <http://ipsnews.net/>.

Amazon Alliance (2006), 'Petrobras Desiste de Construer Estrada no Yasuní', 25 April 2006, <http://Amazonia.org.br> (Washington, DC: Amazon Alliance).

Bebbington, A. (2004), 'Movements and Modernizations, Markets and Municipalities', in Peet, R. and Watts, M. (eds).

Blomberg, R. (1957), *The Naked Aucas. An Account of the Indians of Ecuador* (New Jersey: Essential Books, Inc.).

Brosius, J.P. (1997a), 'Endangered Forest, Endangered People: Environmentalist Representations of Indigenous Knowledge', *Human Ecology* 25: 1, 47–69.

Brosius, J.P. (1997b), 'Prior Transcripts, Divergent Paths: Resistance and Acquiescence to Logging in Sarawak, East Malaysia', *Comparative Studies in Society and History* 39: 3, 468–510.

Castree, N. (2004), 'Differential Geographies: Place, Indigenous Rights and "Local" Resources', *Political Geography* 23, 133–167.

Conklin, B.A. and Graham, L.R. (1995), 'The Shifting Middle Ground: Amazonian Indians and Eco-Politics', *American Anthropologist* 97: 4, 695–710.

Cronon, W. (1996), *Uncommon Ground: Rethinking the Human Place in Nature* (New York: W.W. Norton and Co).

Descola, P. and Palsson, G. (eds) (1996), *Nature and Society: Anthropological Perspectives* (London: Routledge).

El Comercio (1999), 'Proteger la Naturaleza. Yacimientos Petroliferos Fiscales (YPF)', 4 April 1999 (Quito: El Comercio).

—— (2005), 'Los Huaorani Rompen los Acuerdos con Petrobras', 1 July 2005 (Quito: El Comercio) (Washington, DC: Amazon Alliance).

—— (2006a), 'Los Sabios Huao, a Favor de la Selva,' 11 Jan 2006 (Quito: El Comercio) (Washington, DC: Amazon Alliance).

—— (2006b), 'The Division of the Waorani Worsens. La División de los Wao se Acentúa Más' (Quito: El Comercio) (published online 6 April 2006) <www.saveamericasforests.org/yasuni/news>.

—— (2006c), 'The Violence in Yasuní Increases. La Violencia Crece en el Yasuní' (Quito: El Comercio) (published online 29 April 2006) <www.saveamericasforests.org/yasuni/news>.

—— (2006d), 'The Wao Are Exposed on Too Many Fronts. Los Wao se Abren Demasiados Frentes' (Quito: El Comercio) (published online 3 May 2006) <www.saveamericasforests.org/yasuni/news>.

Gavigan, B., Hanon, J., and Ewing, B. (2006), *End of the Spear* (Beverly Hills, CA: Twentieth Century Fox Film Corporation and Every Tribe Entertainment).

Gordon-Warren, P. and Curl, S. (eds) (1981), *Ecuador: In the Shadow of the Volcanoes* (Quito: Ediciones Libri Mundi).

Hames, R.B. and Vickers, W.T. (1983), *Adaptive Responses of Native Amazonians* (New York: Academic Press).

Kane, J. (1995), *Savages* (New York: Alfred A. Knopf).

Kimerling, J. (1991), *Amazon Crude* (New York: Natural Resources Defense Council).

Kingsland, R. (1980), *A Saint Among Savages* (London: Collins).

Laurie, N., Andolina, R., and Radcliffe, S. (2005), 'Ethnodevelopment: Social Movements, Creating Experts and Professionalising Indigenous Knowledge in Ecuador', *Antipode* 37: 3, 470–496.

Li, T. (2000), 'Articulating Indigenous Identity in Indonesia: Resource Politics and the Tribal Slot', *Comparative Studies in Society and History* 42: 1, 149–180.

—— (2004), 'Environment, Indigeneity and Transnationalism', in Peet, R. and Watts, M. (eds).

Lu, F.E. (1999), *Changes in Subsistence Patterns and Resource Use of the Huaorani, Indians in the Ecuadorian Amazon*, Dissertation (Chapel Hill, NC: University of North Carolina at Chapel Hill).

Lucas, K. (2006), 'Ecuador: Logging Activity forms Backdrop to Conflict Between Indians', Inter-Press Service News Agency (published on-line 3 June 2006) <http://ipsnews.net/>.

Mena, P.V., Stallings, J.R., Regalado, J. and Cueva, R. (2000), 'The Sustainability of Current Hunting Practices by the Huaorani' in J. Robinson and E. Bennett (eds).

Peet, R. and Watts, M. (eds) (2004), *Liberation Ecologies. Environment, Development, Social Movements*, 2nd edition (London: Routledge).

Perreault, T. (2003), '"A People With Our Own Identity": Toward a Cultural Politics of Development in Ecuadorian Amazonia', *Environment and Planning D: Society and Space* 21, 583–606.

Proyecto PETRAMAZ (1999), 'Zonas Intangibles de la Amazonia Ecuatoriana. Por la Diversidad Cultural y Biologica' (Quito: Ministerio de Medio Ambiente, Union Europea).

Rainforest Action Network (1997), 'Isolated Huaorani Indians Threatened by US Based Occidental Petroleum Corporation' (San Francisco: Rainforest Action Network) <http://www.ran.org/ran/ran_campaigns/amazonia/huaorani.html>, accessed 9 February 1999.

Ramos, A.R. (1992), 'The Hyperreal Indian', *Serie Antropologia* 135, 1–27.

Reyes, A. and Villavicencio, F. (1999), 'Tagaeri', *Terra Incognita* 1: 3, 13–15.

Rival, L. (1992), 'Social Transformations and the Impact of Formal Schooling on the Huaorani of Amazonian Ecuador', PhD Dissertation (London: University of London).

Rival, L. (1994), 'The Growth of Family Trees: Understanding Huaorani Perceptions of the Forest' *Man* 28, 635–652.

Rival, L. (1996), 'Blowpipes and Spears. The Social Significance of Huaorani Technological Choices' in P. Descola and G. Palsson (eds).

Rival, L. (1997), 'Modernity and the Politics of Identity in an Amazonian Society', *Bulliten of Latin American Research* 16: 2, 137–151.

Rival, L. (1998), *Hijos del Sol, Padres del Jaguar. Los Huaorani de Ayer y Hoy* (Quito: Ediciones Abya Yala).

Rival, L. (2002), *Trekking Through History. The Huaorani of Amazonian Ecuador* (New York: Colombia University Press).

Robarchek, C. and C. (1998), *Waorani. The Contexts of Violence and War* (Fort Worth: Harcourt Brace College Publishing).

Robinson, J. and Bennett, E. (eds) (2000), *Hunting for Sustainability in Tropical Forests* (New York: Columbia University Press).

Sawyer, S. (1997), 'The Politics of Petroleum: Indigenous Contestation of Multinational Oil Development in the Ecuadorian Amazon', MacArthur Consortium Working Paper Series, Institute of International Studies, University of Minnesota.

Sawyer, S. (2004), *Crude Chronicles. Indigenous Politics, Multinational Oil, and Neoliberalism in Ecuador* (Durham, NC: Duke University Press).

Sierra, R., Rodriguez, F. and Losos, E. (1999), 'Forest Resource Use Change During Early Market Integration in Tropical Rain Forests: The Huaorani of Upper Amazonia', *Ecological Economics* 30, 107–119.

Servindi (2006), 'ONU: Denuncian a Petrobras por Violar Derechos de Pueblos Wuaorani (Huaorani)', 18 May 2006 (Washington, DC: Amazon Alliance).

Slater, C. (1995), 'Amazonia as Edenic Narrative', in William Cronon (ed.).

Smith, R. (1993), *Drama Bajo el Manto Amazonico: El Turismo y Otros Problemas de los Huaorani en la Actualidad* [Crisis Under the Canopy: Tourism and Other Problems Facing the Present Day Huaorani] (Quito, Ecuador: Abya-Yala).

SUBIR (1999), 'SUBIR. Sustainable Uses for Biological Resources' (Quito: The SUBIR Project/CARE – Ecuador).

Taussig, M. (1986), *Shamanism, Colonialism, and the Wild Man. A Study in Terror and Healing* (Chicago: University of Chicago Press).

Ulloa, A. (2005), *The Ecological Native: Indigenous Peoples' Movements and Eco-governmentality in Colombia* (New York: Routledge).

Valdivia, G. (2005), 'On Indigeneity, Change, and Representation in the Northeastern Ecuadorian Amazon', *Environment and Planning A* 37, 285–303.

Whitten, N. (ed.) (1981), *Cultural Transformations and Ethnicity in Modern Ecuador* (Urbana: University of Illinois Press).

Walker, C. (1996), *Trinkets and Beads* (New York: First Run/Icarus Films, 52min).

Wallis, E. (1960), *The Dayuma Story; Life Under Auca Spears* (New York: Harper & Row).

Watts, M.J. (1999), 'Petro-Violence. Some Thoughts on Community, Extraction, and Political Ecology', University of California, Berkeley, Institute of International Studies Workshop on Environmental Politics Working Paper, 99–1.

Weatherford, J. (1994), *Savages and Civilization. Who Will Survive* (New York: Crown Publishers, Inc.).

Yost, J.A. (1981a), 'People of the Forest: The Waorani', in Gordon-Warren, P. and Curl, S (eds).

Yost, J.A. (1981b), 'Twenty Years of Contact: The Mechanisms of Change in Wao ("Auca") Culture', in Whitten, N. (ed.).

Yost, J.A. and Kelley, P.M. (1983), 'Shotguns, Blowguns, and Spears: An Analysis of Technological Efficiency', in, Hames, R.B. and Vickers, W.T. (eds).

PART 3
Environmental Movements:
Contested (Re)Scaling of Knowledges, Problems and Narratives

Chapter 7

Confronting Invisibility: Reconstructing Scale in California's Pesticide Drift Conflict

Jill Harrison

Introduction: Pesticide Drift as a Question of Scale

Imagine: You and your family are enjoying a summer evening in your yard when something goes terribly wrong. Suddenly, you have difficulty breathing, your eyes and lungs burn, your youngest child struggles for air, and your other child falls to the ground, vomiting and crying. You smell a strange odour in the air and rush everyone inside. You phone for help and are instructed to stay inside your house and close the windows. You call the neighbours, who report similar symptoms but note that they were told to open all of their windows to air out the house. The symptoms return in waves throughout the night, and when the fire department arrives hours later they are unable to determine the cause of illness. You try to leave the neighbourhood, but police have set up blockades to contain the unknown contaminant. They tell you to relax, to not panic. Several days later you learn that a pesticide being applied on a farm one quarter of a mile away drifted into your neighbourhood, causing over two hundred people to become ill, in some cases hospitalized. You hear from activists that the active chemicals in the spray cause neurological damage, respiratory damage and birth defects. In the weeks and months that follow, you receive no contact from the pesticide applicator or regulatory officials. You and your neighbours are left with fear, lingering illnesses, and hospital bills you cannot afford to pay.

This story characterizes an instance of agricultural *pesticide drift*, a term that refers to the offsite, airborne movement of pesticides away from their target location. Pesticide drift has become an increasingly frequent and controversial issue at the urban-agriculture interface, particularly in the wake of the large-scale drift incidents that have occurred every year or two since 1999 in California's San Joaquin Valley (see Table 7.1 and Map 7.1).[1] In each of these large-scale incidents, up to several hundred workers and/or residents of farmworker communities have been exposed to highly toxic airborne soil fumigants and/or aerially applied insecticides. Pesticide drift typically causes serious acute illness (nausea/vomiting, eye/skin irritation, difficulty breathing) and likely contributes to many chronic diseases, including asthma

---

1 The San Joaquin Valley constitutes the southern end of California's Central Valley and includes all or part of each of the following eight counties: Kings County, Fresno, Kern, Merced, Stanislaus, Madera, San Luis Obispo, and Tulare.

and other lung diseases, cancer, birth defects, immune system suppression, behavioural disorders, and neurological disorders (Kegley et al. 2003; O'Malley 2004).

Table 7.1   Selected major pesticide drift incidents in California's San Joaquin Valley

Date	Town/ Location	County	# People Affected	Pesticide
November 1999	Earlimart	Tulare	170 residents	Metam Sodium
June 2000	Terra Bella	Tulare	24 workers	Chlorpyrifos
June 2002	Arvin	Kern	138 workers	Metam Sodium
June 2002	Arvin	Kern	273 workers and residents	Metam Sodium
October 2003	Lamont	Kern	163 residents + 3 workers	Chloropicrin
May 2004	Arvin	Kern	122 workers	Methamidophos
2005		Kern	at least 42 workers	Metam Sodium
May 2005	Arvin	Kern	27 workers + 6 emerg.crew	Cyfluthrin + Spinosad

Source: DPR Pesticide Illness Surveillance Program Summary Reports (DPR 2006b).

That pesticide drift incidents occur with such regularity in this region is in many ways unsurprising. California agriculture is notoriously pesticide intensive, and a large percentage of the pesticides used within the state are highly toxic and prone to drifting offsite (Kegley et al. 2003). California agriculture accounts for less than three per cent of all US cropland but 25 per cent of the nation's agricultural pesticide use (DPR 2006a; USDA 2002; US EPA 2004, Tables 3.5 and 4.2). All of the chemicals implicated in the major drift incidents on Table 7.1 are exceptionally toxic to humans: chloropicrin is highly acutely toxic, metam sodium is a probable carcinogen and a suspected reproductive and developmental toxicant, and methamidophos and chlorpyrifos are neurotoxins. Furthermore, the particular geographic and atmospheric conditions (e.g. basin shape, inversion layer, heat) of the San Joaquin Valley trap pesticides and other air pollutants and make them volatilize quickly, thus compounding the state's ability to safely regulate pesticide use in a way that would prevent drift.

While pesticide drift might not be surprising, the issue raises troubling questions about the effectiveness of pesticide regulation and the appropriateness of regulatory response. In spite of the continued occurrence of these highly publicized incidents, the documented exposure of thousands of Californians to highly toxic pesticide drift (DPR 2006b), and research findings that hundreds of thousands of Californians are exposed annually to airborne pesticides at levels exceeding those deemed safe by US Environmental Protection Agency (Lee et al. 2002), regulatory response to the issue has been miniscule: measures have been limited to improvements in emergency response rather than restrictions designed to prevent pesticide drift incidents in the first place. How do regulatory officials justify such a minimal regulatory response in this context?

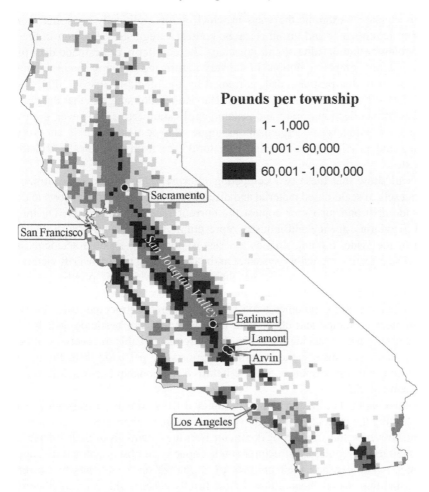

Map 7.1   Map of reported 'Bad Actor' pesticide use in California in
2003 (by township). This map reflects the total number of
pounds of active pesticide ingredients applied per township
(6x6 square mile area); the darkest areas thus reflect the
spaces of highest pesticide use. 'Bad Actors' is a designation
used by Pesticide Action Network to refer to pesticides that
are known or suspected to be highly acutely toxic and/or
capable of causing cancer, reproductive or developmental
disorders, and/or neurological damage (see PAN 2006a)

Source: Department of Pesticide Regulation Pesticide Use Report Data 2003.

In this chapter, I examine the ways in which actors politically construct *scale* in order to legitimize – and, in other cases, contest – regulatory inaction in the face of ongoing pollution, illness, and injustice. The conflict over pesticide drift pivots around highly competing notions of the very nature and extent of the problem. The dominant way that pesticide drift is framed by the media (Harrison 2004) and by regulatory officials and industry leaders (Harrison 2006) suggests that the problem constitutes a collection of rare and isolated application errors. However, a growing group of individuals and organizations argue that pesticide drift is an everyday public health problem that cannot be understood apart from its political and historical context.

I will show that these two competing interpretations of the issue utilize two distinct sets of scale-based material and discursive strategies in their efforts to define pesticide drift and justify or contest the current regulatory response. On the one hand, regulatory agency and industry representatives consistently 'push down' the scale of the issue, framing drift as a series of localized, isolated accidents in an otherwise effective regulatory system. On the other hand, pesticide drift victims and other activists struggle to reconstruct the scale(s) at which the problem is framed and understood in order to reframe the issue as one of everyday risk. I argue that pesticide drift victims and other activists struggle to reconstruct the scale(s) at which the problem is framed and understood in order to release pesticide drift from the localized confines of 'accident', and in so doing make visible the everyday illnesses of marginalized residents, the structural supports for pesticide drift, the need for increased regulatory response, and the iterative relationship between social power and chemical illness.

In this chapter, I utilize the issue of pesticide drift as a timely and revealing case for interrogating the discourses through which environmental problems are understood. I also show how the strength of dominant framings stems from their intersections with ineffective regulatory structures, their capacity to obscure structural supports for ongoing illness, and their abilities to capitalize on – and thus to exacerbate – longstanding social inequalities. I first briefly discuss the relevant theoretical literature on the politics of scale in order to contextualize the notion of scale as a political construct. I then describe the dominant framing of the issue as promoted by pesticide regulatory agency representatives, where the characterization of pesticide drift as a localized series of isolated 'accidents' effectively naturalizes and reinforces the state's minimal regulatory response to the issue. Finally, I turn to the material and discursive processes by which social movement activists strategically manipulate scale in order to illuminate the unofficial, 'everyday' nature of pesticide drift and thus to justify their calls for radical restrictions on the widespread use of chemical pesticides in agricultural production.

This work is based on data that I collected between 2002 and 2006, including over 60 formal in-depth interviews and numerous additional informal qualitative interviews with key actors within the California agriculture regulatory arena and activist community, ethnographic research in the regulatory and activist arenas, and archival research.

Politics of Scale and the Discursive Production of Invisibility

Recent theoretical developments in the politics of scale literature in human geography provide a useful framework for understanding current conflicts over pesticide drift and the relationships between discourse, regulation and power. Interrogations of the politics of scale constitute a rich body of work that challenges the notion of scale as ontologically pre-given and instead investigates 'the ways in which the social construction of scale shapes and is shaped by political and economic processes' (Kurtz 2003: 888). Many contributors have shown that actors engage scales in ways that are advantageous to them, and that the interactions of political economic processes at different scales have real material consequences (Delaney and Leitner 1997; Herod 1997; Marston 2000; Miller 1997; Miller 2000; Smith 1992; Swyngedouw 1997).

This chapter draws more explicitly on the strand of the politics of scale literature that analyzes how actors strategically engage scale-based *discourse* in order to frame an issue in a particular way to effect change, whether to legitimize or to challenge existing power asymmetries (Delaney and Leitner 1997; Kurtz 2003; Miller 1997; Miller 2000; Mitchell 1998; Towers 2000; see also Schmelzkopf and Mulvaney, this volume). This approach is influenced by broader postmodern developments in social science, notably the epistemological understanding that language constructs (not simply reflects) the social world (Fairclough 1995; Foucault 1990 [1978]; Hajer 1995; Jones 1998). My emphasis on power is characteristic of *critical* discourse analysis, which aims to identify the ways in which discourses sustain – but can also be used to subvert and shift – unequal power relations (Fairclough 1995).

This chapter has close affinity with research that illustrates how social movements discursively engage scale in order to justify calls for increased regulatory action (for example, see Kurtz 2003). Here, I build on my previous work on pesticide drift (Harrison 2004, 2006, forthcoming) by focusing on the multiple strategies with which social movement actors leverage scale in order to contest the current regulatory response to pesticide drift. I am particularly concerned with identifying how the material and discursive processes through which actors construct scale pose consequences for social and environmental (in)justice.

Epitomizing industrialized specialty crop production, California agriculture is dependent on high rates of chemical use and also on a ready supply of cheap, skilled labor on a seasonal basis. California's 1.1 million farmworkers are largely politically disenfranchised and socially subordinated: over 90 per cent are immigrants from Mexico, average income falls between $7,500 and $12,500 per year, and anywhere between 50 to 90 per cent are undocumented (NAWS 2002; Rosenberg et al. 1998).[2] The 2000 US Census data similarly indicate that the small, agricultural communities in which large-scale pesticide drift incidents have occurred in recent years (see Table 7.1) on average are poor (17 to 19 per cent of households in Earlimart, Arvin, and Lamont report incomes of less than $10,000 per year) and that the majority of residents are Latino (88 to 90 per cent of the residents in these towns are Hispanic/Latino).

---

2   While survey work suggests that approximately 50 per cent of California's farm labor force is undocumented (or falsely documented), United Farm Worker union insiders suggest that the figure is closer to 90 per cent.

Since at least the late nineteenth century, growers and the state have worked together to actively recruit different ethnic-based migrant farm labour groups; this labour supply has been kept cheap and vulnerable by a combination of physical repression, disparagement, the denial of particular rights, and racist and xenophobic immigration and labour laws that help growers replace one ethnic-based group with another more 'compliant' one in times of labour organizing (often referred to as a 'labor shortage'; McWilliams 1999 [1935]; Wells 1996). However, researchers have noted that the societal propensity to view agricultural landscapes as natural, labour-less spaces compounds farmworkers' relative invisibility (Mitchell 1996; Williams 1973).

Today, border policies deepen immigrants' vulnerabilities by further militarizing the border zone, pushing people to cross in increasingly dangerous and remote deserts and mountains (Nevins 2002). Also, immigration policies and practices exacerbate the historical vulnerability by criminalizing immigrant workers (and their use of social services) much more than the industries that directly and indirectly rely on their labor and vulnerability.[3] Border and immigration policies should be understood in large part as state efforts to deal with the fallout from neoliberal trade policies (such as NAFTA) that increase economic vulnerabilities for many people and thus exacerbate migration patterns (Barry 1995).

In general, although pesticide drift is not solely a farmworker or immigrant problem, these populations likely experience disproportionate exposure, have disproportionately low access to health and legal services, and/or possess a limited ability to make their experiences and concerns visible in local politics.

The Dominant Framing: 'Scaling Down' to 'Accident'

> The system works ... Unfortunately, we have people who don't follow the law (County Agriculture Commissioner, in response to a question about the effectiveness of pesticide drift regulations [*Bakersfield Californian* 2005]).

> It is the power of selection and simplification – or categorization – that gives representations their persuasive power ... They both encourage certain meanings and constrain or limit other meanings ... the rules of social order and the practices of representation go hand in hand (Jones 1998: 27–28).

As I have discussed in greater detail elsewhere (Harrison 2006), the dominant framing of pesticide drift, as reflected in regulatory agency data and statements, portrays pesticide drift as an unfortunate but ultimately small problem. Much of this framing pivots around arguments that pesticide drift incidents are rare. The

---

3   Department of Homeland Security statistics show that of the 206,653 convictions for immigration and naturalization violations between 1993 and 2004, only 436 (amounting to 0.21 per cent) of those violations were for employing undocumented workers (DHS, 2004). A series of government raids in recent years on canneries and other factories employing immigrant workers appears to target this imbalance, yet the extent to which such actions reconfigure burden and responsibility remains unclear and warrants further investigation. Note: the most recent (2005) DHS data do not report employer violations.

Department of Pesticide Regulation's (DPR) Pesticide Illness Surveillance Program (PISP) data show that 1,715 people were classified as suspected or confirmed victims of agricultural pesticide drift from 1998 to 2004 (DPR 2006b).[4] In a PISP summary report, DPR asserted that these figures fairly characterize the scope of the problem: 'DPR maintains a high degree of confidence that the Pesticide Illness Surveillance System captures the majority of agricultural pesticide illnesses ...' (DPR 2006b). Moreover, regulatory officials contextualize these numbers in order to emphasize their insignificance relative to the absolute frequency of pesticide applications. As the then-director of DPR recently stated in a television documentary, 'There are over a million pesticide applications every year [in California]. The incidents of drift are around forty per year, so that's a relatively small number' (quoted in Barbassa 2004a).[5]

Furthermore, officials consistently characterize incidents as 'accidents', as the result of pesticide applicator errors. As one county agriculture commissioner argued, '... in fact it is the people who are not following the rules who are creating the problem'. In a similar vein, a DPR spokesman justified the department's particular response to pesticide drift by noting, 'We don't want the reputation of the industry ruined by a few bad apples or careless acts' (quoted in Barbassa 2005).

In shifting the focus to the circumstances of applicator error and incident response procedures, this framing effectively constricts the scale of analysis to the most local level and particularizes pesticide drift into a series of discrete, isolated events. Framing pesticide drift incidents as being relatively few in number and limited to applicator error enables officials to argue that the system is effective and that no new rules are needed: As a county agriculture commissioner put it, '... when you have a relatively small number of incidents that impact the people or the environment, the system for the most part works'. Consequently, DPR's approach to addressing the problem has been limited to discussing the need for improved emergency response protocol and issuing fines in some cases where errors have been documented.

Structural Supports for 'Accident' Framing

The power of this framing to shape and naturalize regulatory inaction stems in large part from several key structural supports: regulatory structure, regulation language, and longstanding social inequalities. First, the highly *devolved* nature of authority and discretion in California's regulatory structure reinforces this interpretation of pesticide drift as local-level, isolated incidents.[6] The devolution turns problems like

---

4   This number represents 56 per cent of all 3,027 suspected and confirmed agricultural pesticide illnesses during the 1998–2004 time period (DPR 2006b).

5   I should note that in private, confidential conversations, some regulatory scientists deviate from this predominant story, admitting that regulatory programs do a poor job of evaluating numbers of pesticide exposures. I chose to emphasize the narrative of 'accidents' here because they constitute the public 'party line'; I am concerned with institutional response and with the capacity of the predominant narrative to naturalize regulatory failure and thus to legitimize the neglect and exacerbation of environmental illness.

6   This devolution was mandated by the state legislature and designed to allocate discretion to local regulatory officials who could tailor regulatory decisions to local conditions (DPR 2001: 8).

pesticide drift into county-level issues, and pesticide drift incidents are interpreted and treated as localized events rather than part of a more general pattern/trend.

Furthermore, the notion that drift incidents are accidents (caused by errors) resonates rather well in a context where application guidelines are extremely vague, for example, containing language such as 'utilize due care', 'take appropriate safeguards' and 'check for odours'. Pesticide drift events, then, can be easily characterized and shrugged off as accidents if for no other reason than the applicator simply did not 'use due care'.

Finally, the argument that pesticide drift incidents are rare enough to justify only the most minimal regulatory response capitalizes on – and thus exacerbates – the very deeply rooted social inequalities that make California's agricultural abundance possible. Residents of agricultural communities, farm labour representatives, and environmental activists report that drift routinely occurs in agricultural neighbourhoods and workplaces, and that pesticide exposures result in lingering chronic illness, in some cases severely compromising victims' long-term health. In their accounts of pesticide drift, residents and other activists augment official statistics with first-hand, unofficial accounts of pesticide exposures that are either reported but not fully investigated, never reported, or not ever noticed: these are the low-level 'everyday' exposures that permeate the lives of people living in and near agricultural communities. In a recent survey of 321 residents of one farmworking community, 41 per cent of respondents reported that they had been 'drifted on', and more than half of those stated that they had been 'drifted on' two to five times. As one farmworker advocate argued, 'If you ask a room full of farm workers how many of them have been involved in a pesticide drift, at least half will raise their hands' (Vasquez 2005).

Residents and advocates explain that social and political vulnerabilities that plague farmworker communities lead to low levels of reporting and thus limit the accuracy of official pesticide illness data collection systems. Victims and advocates emphasize that poverty and political vulnerability limit farmworkers' access to legal and medical resources, that residents lack knowledge and language skills necessary to pursue their claims, that workers fear job loss or other forms of retaliation from employers, that undocumented victims fear deportation, and that regulators routinely ignore victims' claims or fail to conduct complete investigations. As one community organizer stated, 'There's incidents that are happening all of the time, they just go on not reported. Sometimes the people don't want to get involved – fear of retaliation – or they just simply don't know what it is'. One drift victim noted, 'They said they need sufficient proof. How do you prove it with no money? No money, no doctor!' One farmworker succinctly summarized the dilemma that workers face when exposed to pesticides: 'It's not in their interest to let it be known'.

Moreover, many farmworker advocates argue that poverty and undocumented legal status restrict residents' political clout, resulting in differentially low treatment by regulatory agencies. Residents frequently report that county agriculture commissioners commonly take several days to respond to reports of pesticide drift – by which point the investigations fail to account for the exposure levels that victims actually endured – and rarely conduct follow-up investigations. Consequently, most experiences of 'everyday' pesticide drift are never reflected in official data. To explain such problems, one particularly sympathetic state senator stated, 'there is a thought that these are, if you will, very dispensable folks'.

The dominant framing of pesticide drift is thus one in which the scale of the problem has been manipulated in a way that particularizes pesticide drift into a series of discrete, isolated, and local incidents. Notably, this framing of the issue gains considerable purchase and consequence through its intersections and interactions with key features of regulatory structure and longstanding social inequalities, effectively obscuring – or erasing – everyday illness and naturalizing the state's minimal regulatory response to ongoing pesticide drift.

Social Movement Response

> When 'accidents' like this keep happening, it's no longer an accident but a poorly defined system that guarantees these such poisonings will continue (pesticide drift activist quoted in DeAnda 2006: 2).

> We must make allowance for the complex and unstable process whereby discourse can be both an instrument and an effect of power, but also a hindrance, a stumbling block, a point of resistance and a starting point for an opposing strategy. Discourse transmits and produces power; it reinforces it, but also undermines and exposes it, renders it fragile and makes it possible to thwart it (Foucault 1990 [1978]: 101).

The localized 'accident' narrative is the dominant, but not the only way to tell the story of pesticide drift. A collection of farmworkers, residents of agricultural communities, and many farmworker and environmental advocates that have come to embody what I call a nascent pesticide drift social movement argue that pesticide drift is indeed an everyday problem. The accidents that regulatory officials use to define the problem serve a central but much different purpose for activists: they are the events that capture media attention, compel activism, and serve as a tool for reminding others about the face of injustice – but which are really the outliers in a landscape of *everyday contamination*.[7] As one activist noted, 'It's the day-to-day drift that is the biggest problem, because it's the one that isn't even looked at, not even considered, whereas the accidents get the front page of the newspaper'.

Several key pesticide drift incidents served as catalysts for this social movement. In response to a 1999 pesticide drift incident in Earlimart and several subsequent large-scale pesticide drift events, drift victims and some other outspoken community members began to form community-based organizations in order to address their shared concerns about lingering health problems and inadequate emergency and regulatory response. These community groups soon joined forces with other local, regional, and national social justice and environmental organizations (including Pesticide Action Network, Center for Race Poverty and the Environment, and

---

7   See Sewell (1996) and Cronon (1992) for reflexive discussions of the role of events in historical narratives.

California Rural Legal Assistance Foundation) to address the problem of pesticide drift. In 2002, this grassroots activism coalesced and began to take shape as an emergent nationwide social movement when the San Francisco-based coalition group Californians for Pesticide Reform (CPR) decided to target pesticide drift as its largest campaign, with the southern end of the San Joaquin Valley as its regional focus. CPR pulls together the overlapping work of California's various community and regional pesticide organizations and builds connections with grassroots pesticide drift campaigns in other states nationwide. Since 2002, the activists involved in CPR's pesticide drift campaign have pursued county, state, and federal regulatory reform, state-level policy change, community education and organizing, and health and pollution data collection. Pesticide drift provides an opportunity for these groups to focus on the unrecognized pesticide illness in and near the site of production and also to gain political traction by linking agricultural workplace pesticide exposures with those of nearby residents.

While pesticide exposure is the uniting theme, it is important to note that most of these groups are either focused on farmworker community issues or have a major farmworker justice campaign within their work. Several key pesticide drift activist-leaders developed intellectually through social justice and civil rights work; for example, CPR director David Chatfield was greatly influenced by the United Farm Workers, for which he worked in the tumultuous summers of 1973 and 1974.[8] The environmental justice (EJ) movement's critical attention to the relationships between social inequalities and pollution resonates with many pesticide drift activists, who have built alliances with EJ groups (including the nascent Central California Environmental Justice Network) and continue to play a strong role in shaping DPR's new EJ pilot project in the San Joaquin Valley (DPR 2006c).

Pesticide drift activists argue that, in the context of major structural inequalities facing residents of farmworker communities, regulators' 'accident' discourse obscures a considerable amount of illness and thus naturalizes a regulatory response that fails to adequately protect public health from agricultural pesticide drift. To combat this invisibility, activists strategically *manipulate* scale in two principal ways: by deepening regulatory agencies' local scale characterizations of pesticide drift and simultaneously pushing up the scale of analysis.

Constructing a more complex local

Many of the pesticide drift activist discourses and tactics are designed to confront and deepen the dominant, 'accident' framing of the problem by making visible the ways in which poverty and other forms of social subordination compound the effects of pesticide drift, particularly for disenfranchised farmworker communities. Social movement work pivots around community organizing and education efforts to empower, encourage, and inform farmworker community residents. Activist leaders

---

8   Chatfield also worked during the 1970s for the American Friends Service Committee's (AFSC) rural outreach education program, a project to educate people outside of the Bay Area about AFSC social justice issues; he also helped start AFSC's Rural Education and Action Project, which focused on the relationships between industrial agriculture and rural poverty.

continually work to encourage victims to disseminate their own stories at hearings, meetings, and rallies, in an effort to saturate the regulatory arena with firsthand stories that document the relationships between pesticide illness and longstanding race- and class-based vulnerabilities. Activists back up residents' anecdotal stories about everyday illness through coordinating grassroots air monitoring studies, disseminating those results, and presenting other data that illustrate the widespread nature of pesticide drift.[9]

Victims emphasize the lingering health problems they attribute to pesticide exposure – asthma, migraine headaches, miscarriages, rashes, learning disabilities, behavioral changes, emotional distress – thereby deepening the understanding of pesticide drift beyond the narrow frame of symptoms experienced at the time of exposure. According to one community organizer, Earlimart residents refer to their health in terms of 'before the accident' and 'after the accident'. Tweaking regulatory officials' propensity to focus on the local scale and isolated incidents, this combination of tactics engages a more complex and damning set of 'local' variables, thereby illustrating the inadequacies in pesticide regulatory practice.

Such tactics suggest the need for changes in regulatory response that account for the fact that social inequalities inevitably limit the scope and effectiveness of regulatory programs. Activists thus call for a widespread implementation of the precautionary principle in pesticide risk assessment, policy, and regulation including the following possibilities:

- restrict the use of the most-toxic and drift-prone chemicals and application methods,
- phase out chemicals for which less-toxic alternatives exist,
- standardize pesticide regulations across all counties,
- make health-protective (rather than industry-protective) decisions in cases of data gaps or other forms of scientific/regulatory uncertainty,
- replace pesticide regulatory upper management (who typically hail from industry) with public health leaders,
- aggressively prosecute violators,
- and increase funding for alternative (less-toxic) pest control technologies and methods.

'Pushing up' the scale of analysis

While focusing a considerable amount of effort on organizing in farmworker communities and making visible the disproportionate burden of pesticide illness faced by those populations, activists also recognize that victims, concerned residents, and other activists are not limited to farmworkers. They also recognize that the large-

---

9   For initial results of Pesticide Action Network's Drift Catcher air monitoring project, see PAN (2006b). For additional data, see, in particular, Kegley et al. (2003). Lee et al.'s (2002) analysis of air monitoring data, pesticide use data, and Census population data 'suggest a potential for exposures and risks, similar to those calculated in this risk assessment, for hundreds of thousands of people in California' (Lee et al. 2002: 1181).

scale incidents that disproportionately affect farmworker communities are not the only forms of pesticide drift: pesticides contribute to many broadly recognized (and federally regulated) forms of air pollution, notably particulate matter and ozone. Movement leaders thus emphasize the need to frame pesticide drift as *air pollution* to emphasize the nature of pesticide drift as a broad-scale, diffuse, 'everyday' problem that *transgresses* local jurisdictional boundaries such as counties. As one activist leader stated at the 2005 CPR conference, 'CPR's goal is to show that pesticides and air pollution are connected'. Stretching the framing of the problem in this way has enabled pesticide drift activists to capitalize on the political momentum and muscle of the growing social movement of (more politically enfranchised) Californians dedicated to tackling the San Joaquin Valley's increasingly notorious air pollution problems. As a result, air pollution activists now constitute an increasingly powerful multi-lingual and multi-cultural force (see Figure 7.1).[10]

Figure 7.1   Braulio Martinez, of Alpaugh, California, holds a sign during a rally outside of a public hearing to call for stronger, health-protective measures against pesticide-related smog at the Kearney Agricultural Research Center in Parlier, California. The sign reads, 'We don't want any more contamination'. Photo taken 14 August 2006 in Parlier, California, by Teresa Douglass (Visalia Times-Delta)

---

10 The San Joaquin Valley has some of the worst air quality in the nation; it has fallen into the status of 'severe non-compliance' of Clean Air Act standards and contains four of the nation's top five most ozone-polluted cities and three of the nation's top five metropolitan areas most polluted by year-round particle pollution (annual PM2.5) (ALA 2005).

As one example of activists' efforts to insert pesticides into current air pollution debates, a disparate coalition of groups across California brought a successful lawsuit against the Department of Pesticide Regulation for its failure to reduce the contribution of pesticides to the overall load of volatile organic compounds (VOCs) – a primary ingredient in the formation of ground-level ozone.[11] Activists note the role of pesticides and the subsequent air pollutants to which they contribute (such as ozone) in exacerbating lung diseases, including the San Joaquin Valley's disproportionately high rates of childhood asthma. Such points illustrate the long-term and geographically dispersed nature of pesticide illness and the broad array of social costs associated with regional air pollution. According to one study, ozone and particulate matter air pollution cost the San Joaquin Valley over \$3 billion per year in premature deaths, chronic and acute bronchitis, lost work days, hospital admissions, asthma attacks and school absences (Hall et al. 2006).

Furthermore, framing pesticide drift as air pollution strategically broadens the scale at which pesticide drift is understood, thereby releasing the problem from the discursive constrictions of regulatory officials' 'accident' framing and highlighting the need for statewide or federal regulatory response. CPR leaders emphasize the importance of addressing the issue at the statewide level: 'state level work is the place where we can seek to make real changes that reduce the exposure of communities to drift' (CPR 2003). Social movement publications lament the wide variations in pesticide use guidelines across counties and emphasize that all Californians deserve equal protections from pesticide drift (PAN 2004).

Conclusion

In this chapter, I have discussed the political conflict over agricultural pesticide drift in California as the basis for interrogating the political construction of scale and the ways in which interactions between scale and power serve to erase and obscure environmental injustices. While regulatory officials' 'downscaled' framing of pesticide drift legitimizes and naturalizes regulatory inaction, pesticide drift activists strategically manipulate the framing of the problem in multiscalar ways. Activists deepen the local-scale analysis of pesticide drift, making visible the longstanding race- and class-based inequalities that obscure many pesticide illnesses. At the same time, activists struggle to reframe the problem of pesticide drift as air pollution, capitalizing on a broader base of political support, 'pushing up' the scale at which the problem is understood, and justifying calls for higher-scale regulatory reform to protect public health from the adverse effects of pesticide drift. This work illustrates the inherently political and socially constructed nature of scale, as well as the ability

---

11 The environmental law firm Center on Race Poverty and the Environment represented a variety of pesticide activist and air pollution groups (including Association of Irritated Residents, Communities and Children's Advocates Against Pesticide Poisoning, the Wishtoyo Foundation, Ventura CoastKeeper, and El Comité para el Bienestar de Earlimart) when it filed a citizen's suit against the Department of Pesticide Regulation and the Air Resources Board in May 2004 because of the state agencies' failure to reduce pesticides' contributions to the Central Valley's ozone problems. A federal court found in favor of the groups in May 2006.

of scalar constructions to capitalize on – and thus to reinforce, if not also exacerbate – longstanding social inequalities, inadequate regulatory structures, and ongoing chemical illness caused by the everyday 'accidents' of pesticide drift.

Acknowledgements

This research was supported in part by generous funding from the University of California Institute for Labor and Employment, the UC Santa Cruz Environmental Studies Department, and the UC Santa Cruz Graduate Division. Special thanks to Max Boykoff, Kyle Evered, David Goodman, Michael Goodman, Julie Guthman, and Dustin Mulvaney for their careful and encouraging feedback on this chapter.

References

ALA (2005), 'Metropolitan Areas Most Polluted by Year-Round Particle Pollution, and 25 Most Ozone-Polluted Cities (Annual PM2.5)', American Lung Association State of the Air 2005.

Bakersfield Californian (2005), 'Players Discuss Current System for Pesticide Regulation', 17 September.

Barbassa, J. (2004a), 'Little Help for Workers Exposed to Pesticides: 19 Recently Sickened in Drift Incident', *San Diego Union-Tribune*, 22 May.

Barbassa, J. (2005), 'Activists Say Large Settlements, Fines, Send Message to Pesticide Applicators', *San Diego Tribune*, 23 November.

Barry, T. (1995), *Free Trade and the Farm Crisis in Mexico* (Boston: South End Press).

Chavez, L.R. (2001), *Covering Immigration: Popular Images and the Politics of the Nation* (Berkeley: University of California Press).

Cox, K.R. (1997), *Spaces of Globalization: Reasserting the Power of the Local* (New York: Guilford Press).

CPR (Californians for Pesticide Reform) (2003), 'Californians for Pesticide Reform', Policy Goal Memo, 1 August.

Cronon, W. (1992), 'A Place for Stories: Nature, History, and Narrative', *Journal of American History* 78, 1347–76.

DeAnda, T. (2006), 'Community Members Compensated for Pesticide Drift Poisoning: Three Years Later, Arvin Residents Win Settlement Against Applicator, Farmer', *CPR Resource* No. 16, March 2006.

Delaney, D. and Leitner, H. (1997), 'The Political Construction of Scale', *Political Geography* 16, 93–7.

DHS (Department of Homeland Security) (2004), 'Yearbook of Immigration Statistics 2004, Table 50: "Convictions for Immigration, Naturalization, and Other Violations: Fiscal Years 1993–2004"', <http:// www.dhs.gov/ximgtn/statistics/ publications/YrBk04En.shtm>, accessed 20 February 2007.

DPR (Department of Pesticide Regulation) (2001), *Regulating Pesticides: The California Story* (Sacramento, CA: Department of Pesticide Regulation).

DPR (Department of Pesticide Regulation) (2006a), 'Summary of Pesticide Use Report Data 2004, Indexed by Commodity' (Sacramento, CA: Department of Pesticide Regulation).

DPR (Department of Pesticide Regulation) (2006b), 'Pesticide Illness Surveillance Program Summary Reports' (Sacramento, CA: Department of Pesticide Regulation).

DPR (Department of Pesticide Regulation) (2006c), 'DPR Environmental Justice Program' [webpage] <http://www.cdpr.ca.gov/docs/envjust/pilot_proj/index.htm>.

DuPuis, E.M. (2004), *Smoke and Mirrors: The Politics and Culture of Air Pollution* (New York: New York University Press).

Fairclough, N. (1995), *Critical Discourse Analysis: The Critical Study of Language* (London: Longman).

Foucault, M. (1990 [1978]), *The History of Sexuality Volume 1: An Introduction* (New York: Vintage Books).

Hall, J.V., Brajer, V., and Lurmann, F.W. (2006), 'The Health and Related Economic Benefits of Attaining Healthful Air in the San Joaquin Valley', Institute for Economic and Environmental Studies (Fullerton, CA: California State University).

Hajer, M.A. (1995), *The Politics of Environmental Discourse: Ecological Modernization and the Policy Process* (Oxford: Clarendon).

Harrison, J. (2004), 'Invisible People, Invisible Places: Connecting Air Pollution and Pesticide Drift in California', in DuPuis (ed.).

Harrison, J. (2006), '"Accidents" and Invisibilities: Scaled Discourse and the Naturalization of Regulatory Neglect in California's Pesticide Drift Conflict', *Political Geography* 25, 506–529.

Harrison, J. (forthcoming), 'Abandoned Bodies and Spaces of Sacrifice: Pesticide Drift Activism and the Contestation of Neoliberal Environmental Politics in California', *Geoforum*.

Herod, A. (1997), 'Labor's Social Praxis and the Geography of Contract Bargaining in the US East Coast Longshore Industry, 1953–89', *Political Geography* 16, 145–169.

Herod, A. (ed.) (1998), *Organizing the Landscape: Geographical Perspectives on Labor Unionism* (Minneapolis: University of Minnesota Press).

Jones, K.T. (1998), 'Scale as Epistemology', *Political Geography* 17, 25–28.

Kurtz, H.E. (2003), 'Scale Frames and Counter-Scale Frames: Constructing the Problem of Environmental Injustice', *Political Geography* 22, 887–916.

LaDou, J. (ed.), *Current Occupational and Environmental Medicine* (New York: Lange Medical Books/McGraw-Hill).

Lee, S., McLaughlin, R., Harnly, M., Gunier, R., and Kreutzer, R. (2002), 'Community Exposures to Airborne Agricultural Pesticides in California: Ranking of Inhalation Risks', *Environmental Health Perspectives* 110: 12, 1175–1184.

Marston, S. (2000), 'The Social Construction of Scale', *Progress in Human Geography* 24, 219–242.

McWilliams, C. (1999 [1935]), *Factories in the Field: The Story of Migratory Farm Labor in California* (Berkeley: University of California Press).

Miller, B. (1997), 'Political Action and the Geography of Defense Investment: Geographical Scale and the Representation of the Massachusetts Miracle', *Political Geography* 16: 2, 171–85.

Miller, B.A. (2000), *Geography and Social Movements: Comparing Antinuclear Activism in the Boston Area* (Minneapolis: University of Minnesota Press).

Mitchell, D. (1996), *The Lie of the Land: Migrant Workers and the California Landscape* (Minneapolis, University of Minnesota Press).

Mitchell, D. (1998), 'The Scales of Justice: Localist Ideology, Large-Scale Production, and Agricultural Labor's Geography of Resistance in 1930s California', in Herod (ed.).

NAWS (National Agricultural Workers Survey) (2002), 'Demographic and Employment Profile of United States Farm Workers', <http://www.dol.gov/asp/programs/agworker/report9/chapter6.htm#summary>.

Nevins, J. (2002), *Operation Gatekeeper: The Rise of the 'Illegal Alien' and the Making of the US-Mexico Boundary* (New York: Routledge).

O'Malley, M. (2004), 'Pesticides', in LaDou (ed.).

PAN (Pesticide Action Network) (2004), 'Pesticide Drift Sickens Residents', 16 January 2004, PANUPS.

PAN (Pesticide Action Network) (2006a), 'About the data', PAN Pesticide Database [website], <http://www.pesticideinfo.org/Docs/data.html>.

PAN. (2006b), 'Drift Catcher Results' [website], <http://www.panna.org/campaigns/driftCatcherResults.html>.

Rosenberg, H., Steirman, A., Gabbard, S. and Mines, R. (1998), 'Who Works on California Farms?', NAWHS Report No. 7 (Washington, DC: US Department of Labor, Office of the Assistant Secretary for Policy).

Sewell, E.H. (1996), 'Historical Events as Transformations of Structures: Inventing Revolution at the Bastille', *Theory and Society* 25, 841–881.

Smith, N. (1992), 'Contours of a Spatialized Politics: Homeless Vehicles and the Production of Geographic Scale', *Social Text* 33, 55–81.

Swyngedouw, E. (1997), 'Neither Global nor Local: "Glocalization" and the Politics of Scale', in Cox (ed.).

Towers, G. (2000), 'Applying the Political Geography of Scale: Grassroots Strategies and Environmental Justice', *Professional Geographer* 52, 23–36.

USDA (US Department of Food and Agriculture) (2002), 'Chapter 1: California State Level Data', *Census of Agriculture*, Volume 1, <http://www.nass.usda.gov/census/census02/volume1/ca/index1.htm>, accessed 26 March 2006.

US EPA (US Environmental Protection Agency) (2004), 'Pesticides Industry Sales and Usage 2000 and 2001 Market Estimates' (Washington, DC: Office of Prevention, Pesticides, and Toxic Substances), <http://www.epa.gov/oppbead1/pestsales/01pestsales/table_of_contents2001.html>.

Vasquez, R. (2005), 'Trouble in the Air: Dozens Feel Ill after Toxic Gas Drifts into Neighborhood', *Monterey County Weekly*, 13 October.

Wells, M. (1996), *Strawberry Fields: Politics, Class, and Work in California Agriculture* (Ithaca: Cornell University Press).

Williams, R. (1973), *The Country and the City* (New York: Oxford).

Chapter 8

Scale and Narrative in the Struggle for Environment and Livelihood in Vieques, Puerto Rico

Karen Schmelzkopf

Introduction

Claudio Minca reminds us that 'geographical representations not only narrate reality but also contribute to its construction' (2001: 196). For instance, here is one geographical representation of Vieques, Puerto Rico:

> About 7 miles (11km) east of the big island of Puerto Rico lies Vieques (Bee-*ay*-kase), an island about twice as large as New York's Manhattan, with about 9,300 inhabitants and some 40 palm-lined white-sand beaches. From World War II until 2003, about two-thirds of the 21-mile (34km) long island was controlled by US military forces. Much of the government-owned land is now leased for cattle grazing (*Frommer's*, no date).

Here is another geographical representation:

> The island is located between Puerto Rico and St. Thomas (USVI), and is separated from the southeast coast of Puerto Rico by approximately 8 miles of sea, although if you are taking a ferry from Fajardo the distance will be 18 miles (*Welcome to Puerto Rico*, no date).

I suggest a third representation:

> Vieques has been determined in large part by external interests in its strategic geographic situation, more often that not at the expense of the site characteristics of its people and environment. In the 1840s the island was conquered by the Spanish, who annexed it to Puerto Rico and sold off all farmable land to five wealthy landowners. In 1898, after the Spanish American War, Vieques along with the rest of Puerto Rico was colonized by the United States. In 1941 the US Navy appropriated 27,000 of Vieques's 33,000 acres to be used as a military training and testing site. Residents were squeezed into an 11,000-acre strip of Navy-owned land in the middle of the island between the bombing range and an ammunition storage depot. The residents lost their agricultural land, and the environment was severely contaminated. In 1999, after the death of a civilian, Viequenses reached out to the rest of the world to join with them in their opposition of the Navy occupation. In 2003, as a result of these protests, the Navy closed the base.

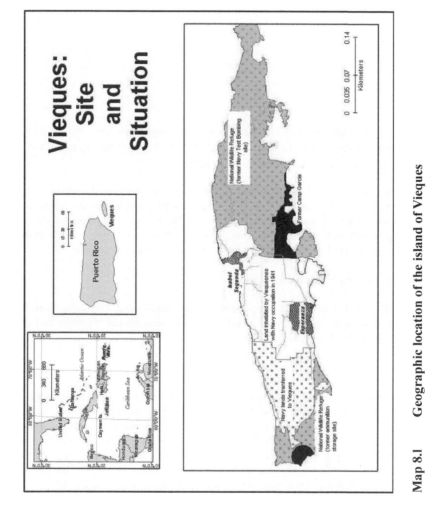

Map 8.1   Geographic location of the island of Vieques

All three representations *narrate* reality: Vieques is a small island with a particular latitude and longitude and relative location, influenced by discourses of strategy, national security, colonialism, capital, leisure, citizenship, the right to place, and so on. All three representations *construct* reality: they are mappings with generalizations, abstractions, simplifications, and omissions. The first two narratives are geographical representations of site and situation; my third narrative describes power and the politics of scale. All three are crucial within the conflict over Vieques.

In this chapter I develop a more detailed geographic representation of Vieques. I use the conceptual apparatus of narrative and scale in order to frame the conflicts over space, environment, and livelihood on the island. I argue that both the Navy and the anti-Navy protestors used what Condit (1987) calls 'rhetorical narratives': accounts of real or imagined events told in order to persuade people.

Stott and Sullivan (2000: 1–2) argue that political ecology is concerned with 'identifying power relationships supported by such narratives' in order to understand the consequences 'for economic and social development, and particularly for constraining possibilities for self-determination'. They point to two questions that are critical to this study: 'Who decides the conditions of truth?' and 'Who has the right to decide for society?'

In the contest over Vieques, these two questions have been played out as a politics of scale, where scale is constructed to define a legitimacy to the 'place' (Herod and Wright 2002) that is Vieques. Purcell and Brown (2005: 608) note that, 'scale and scalar configurations are not independent variables that can cause outcomes, rather they are a strategy used by political groups to pursue a particular agenda.' For example, within a narrative space at one scale of analysis, national security seems to trump everything in Vieques. By constructing scale through this national and now global security perspective, the Navy effectively erased Vieques as a place. On the other hand, at the scale of the local the land belongs to the Vieques people. Yet, the Viequense anti-Navy protestors had to expand beyond the 'local' to actively frame their narrative at the perspectives of the regional, the national, and the global scale to become visible to a wider audience.

I suggest that Vieques is an example of what Swyngedouw calls a 'scaled' place: 'the embodiment of social relations of empowerment and disempowerment' (1997: 169). Moreover, in important ways Vieques has been *defined* by scale. Scale is very porous for Vieques, with fluid border and identities. Since the Spanish colonization, the environment and the livelihood of the people has been dependent on the external interests of either sugar plantation owners or the US military. Ethnicity for Viequenses, people who were born or whose parents were born on Vieques, reflects their history: pre-Columbian settlement by Taino Indians, Spanish colonists, French and Dutch planters, free blacks and slaves, and large-scale immigration from other Caribbean and Latin American countries. Politically, the island is a municipality of Puerto Rico, which in turn is a colony of the US; and as Lloréns explains: 'As United States citizens, the only border Puerto Ricans cross is the imaginary border, which greets us at US airports in another language' (2006: 78). Moreover, because Puerto Ricans are a translocal community, going back and forth between the islands and the US mainland (Duany 2002: 32), many Viequenses spend their lives on the US mainland, never visiting Vieques or else living there only sporadically.

Underlying this porousness is restriction and subordination. While there are no customs agents or visa requirements when Puerto Ricans cross into the US mainland, the effects of these 'imaginary' borders are quite real and are permeable only in accordance with the colonial restrictions set up by the US. Their statutory citizenship can be revoked at any time. When living in Puerto Rico they cannot vote for US president and they have no representation in the US Congress. Nevertheless, when there was conscription, Puerto Rican men were drafted into the US military.

Generations of Viequenses have lived within a discursive regime described by Torres as 'the legacy of colonized perceptual and cognitive behaviour, expressed in a belief system of powerlessness that prevented individual and collective action' (2005: 11). Powerlessness presupposes hegemonic power structures and since the 16th century this power has been wielded by colonists, by plantation owners, and by the US government – particularly the US Navy. Yet, while very few Viequenses have ever owned their land, McCaffrey argues that the Viequenses have an identity based on being excluded from the land (2002: 123). It is this – an emotional longing as well as the economic need for their environment – that has motivated resistance against this hegemony of the US military.

I will show that although Viequenses were marginalized and their legal claims to place were tenuous, they promoted a powerful discourse within a conducive social and political climate that allowed them to *use* their fluid borders to jump scale and defeat the institutional power of the US Navy.

Vieques and the Navy

Vieques is the only inhabited US island ever to have a bombing range, and the Navy insisted it was unique, 'the crown jewel of training areas ... the only place on the East Coast where aircraft, naval surface ships and ground forces could employ combined arms training with live ammunition under realistic conditions' (Johnson and Jones 1999: A21). Although they paid no fees or taxes to Vieques, the Navy had full control of the land, the water and air routes, aquifers, and held the title to the resettlement tracts where residents were relocated (McCaffrey 2002: 3).

Prior to the occupation by the Navy, most Viequenses lived and worked on sugar plantations. Although they did not own the land, they did have usufruct rights to farm the land, to gather coconuts, fruit, and wood, and to fish in the ocean and the mangroves. When the Navy came in 1941, they took over two-thirds of the land, forced Viequenses from their homes and into the middle of the island where the soil was less fertile, and cut off their usufruct rights and their access to most coastal waters, the mangroves, and the areas where they had collected food and wood (McCaffrey 2006: 129).

The Navy discouraged economic development and tourism, which they viewed as potential competition for land and political power. Depriving the Vieques of almost any type of livelihood meant both an unemployment rate and a poverty rate of over 75 per cent. The Navy also believed that economic marginalization would deter social mobilization and political opposition to their presence (Grusky 1992). In general this was true: as CPRDV (*Comité Pro Rescate y Desarrollo De Vieques*)[1] member Christina Corrado pointed out, for thirty years Viequenses just tried to survive (personal communication, 2002). With no land available for agriculture or cattle grazing, most residents were either forced to migrate for work or support themselves as fishermen (Baver 2006: 104). However, in 1975 protestors in neighboring Culebra succeeded in getting their naval operations shut down and the Navy transferred the operations to Vieques. The subsequent increase in bombing activity decimated the fish catch. In response, the Vieques fishermen came together in what was known as the Fishermen's Wars, and blocked training and bombing activity (McCaffrey 2002: 77–79). The protest was successful and the result was a 1983 Memorandum of Understanding in which the Navy acknowledged they had a responsibility toward Vieques.

Under the auspices of the US Department of Defense, the Memorandum focused on four issues: improvement of the welfare of the residents; beneficial use of Navy lands; limitations on use of ordnance; and environmental protection. The Navy solicited only military contracts, and while many projects were planned, including a jewelry manufacturing industrial park, cattle and hay production, and the production of electronics equipment, few came to fruition. By the 1990s the unemployment rate was higher than it had been in 1983 (McCaffrey 2002: 100). It was not surprising when a 1999 report to the Secretary of Defense by the Special Panel on Military Operations on Vieques concluded that

> the constant violations, in both the spirit and the letter, of the 1983 Memorandum of Understanding, by the Navy, together with the limitations inherent therein, have made it a totally ineffective instrument to protect the interests of the population of Vieques, vis-à-vis the activities of the Navy (US Department of Defense, 1999).

In 1993, in response to the formation of a commission by the US government to investigate closing military bases, residents of Vieques formed CPRDV to promote closing the Navy base on Vieques (Baver 2006: 105). Their goal went beyond demilitarization to include decontamination of the land, sustainable development, and the return of the land to the people of Vieques. They had their first opportunity for action within a year, when the Navy announced the construction of Relocatable Over-the-Horizon Radar (ROTHR) on Vieques and in the Lajas Valley on the main island. CPRDV, in short, jumped scale to join forces with protestors from Lajas to fight the radar, arguing that it would eliminate agricultural land and lead to environmental and health problems. Although unsuccessful, it led to one of the largest protests in Puerto Rican (Baver 2006: 105).

In April 1999 a civilian employee of the Navy, Vieques native David Sanes, died after a bomb hit an observation tower instead of its target. CPRDV stormed the Navy's Camp Garcia and set up civil disobedience camps, managing to stop all military operations for a year. They reached out to regional, national, and global organizations for help – further transgressing scale – and after several years of widespread protests, the Navy closed operations in western Vieques in 2001 and except

---

1   Committee for the Rescue and Development of Vieques.

for 100 acres to continue operating the ROTHR, closed the rest of their operations in May 2003. The Navy's congressional allies had vowed the land would never leave federal control if the Navy was forced to leave (Berman Santana 2002), and indeed, they gave almost 18,000 to the US Department of the Interior's Fish and Wildlife Service (FWS), for a National Wildlife Refuge, 800 acres to the Puerto Rico Conservation Trust, and only 4,000 acres on the west to the municipality of Vieques (see Map 8.1).

Politics of Scale and the Narratives of the Navy and Activists

The Navy

During the contest over the fate of the Naval base on Vieques, the Navy and the US government developed a politics of scale in a narrative contrasting the fate of a small island to the need for national security. Republican Senator John Warner stated it this way: 'The safety and security of this great nation, and Puerto Rico – a part of this great nation – is dependent upon the existence of this training for our young men and women and the maintenance of these installations' (US Armed Services Committee 1999). The federal courts made the same point in response to a suit brought by the Vieques Conservation and Historical Trust to stop the bombing and repair all of the environmental damage, saying they '… cannot simply zoom in on the concerns of the United States citizens residing on Vieques, but must pan back and keep the larger picture in focus … national security is too important of an issue to be neglected' (*Vieques Conservation and Historical Trust v. Bush*, n122, cited in Ryder 2004: 443). Within their narrative, the Navy erased Vieques as a place, repositioning it as a useable space.

The Navy's use of narratives of power and scale was evoked from the confident perspective of a utilitarian ethics. They elevated the so-called 'uniqueness' of the island's situation over the rights of the residents, refusing to even consider alternatives until international protests forced their hand. For the most part, Viequenses were comfortably invisible to naval and government officials, geographically distant from Washington and from the nearest officers club on the mainland of Puerto Rico, and lost in the hierarchical rhetoric of military needs and national security. Moreover, as environmental lawyer Robert Kennedy, Jr. emphasized, 'Vieques does not have a congressional representative. They don't have the political avenues that most Americans do to reach the Navy and to reason with them' (CNN 2001).

Since the Navy provided no jobs on Vieques and had not fulfilled the requirements of the Memorandum, they used a narrative of environmentalism to resist change. For instance, the director of the Navy's Environmental Protection, Safety and Occupational Health said, 'Vieques is as unique and beautiful as it is today because of the environmental stewardship of the Navy' (Navy Environmental News 1999). Yet, for sixty-two years, the Navy exploded ordnance made from heavy metal carcinogens, tested chemical and biological weapons in the 1960s, used napalm in 1993, and dropped uranium shells in 1999. Among the detritus left by the Navy are discharged chemicals, missile propellants, metal debris, unexploded ordnance, and at least fourteen inadequately prepared dump sites for solvents, solid waste, and diesel fuel that leached into the soil. There are two known submarine disposal areas, and unexploded ordnance lying on the coral reef and the ocean floor. Contaminants include PCBs, depleted uranium, TNT, perchlorate, lithium, mercury, lead, copper, magnesium, and pesticides., lithium, mercury, lead, copper, magnesium, and pesticides (US Environmental Protection Agency 2004). Research by the Puerto Rico Department of Health showed a 31 per cent higher cancer rate and a 50 per cent higher heart disease than the rest of Puerto Rico, a mortality rate of 10.8 deaths per 1,000 inhabitants, and mercury in Viequense women exceeding allowable levels by 26.8 per cent (Ortiz-Roque and Lopez-Rivera 2004; Wilcox, 2004). In response, the Navy cited a questionable November 2003 report from the Agency for Toxic Substances and Disease Registry that the water is clean, the fish are safe to eat, the levels of chemicals are not harmful, and its activities had no negative impact on residents' health. The Agency, however, had not sampled the rainfall systems used as drinking water, and scattered throughout the report are such phrases as 'the validity and utility of the data is uncertain' (Agency for Toxic Substances and Disease Registry, 2003).

Because the acreage given to the FWS and the Conservation Trust is intended for only 'moderate' use, it will only have 'moderate' cleanup. Further, the US House Armed Services Committee declared the 980-acre former bombing site uninhabitable, which means the Navy will not have to clean it up at all. Ironically, they also gave the site a wilderness designation, though wilderness is defined in the Wilderness Act of 1964 as a place 'where the earth and its community are untrammelled by man … retaining its primeval character and influence … and that generally appears to have been affected primarily by forces of nature, with the imprint of man's work substantially unnoticeable' (Goldberg, 2004).

After vigorous campaigning by CPRDV members, in 2005 the US Environmental Protection Agency designated Vieques a Superfund Site, identifying it is one of the most hazardous sites in the US with priority for cleanup. In a Joint Statement before House Appropriations Committee, however, the Navy portrayed a sense of bewilderment at this designation, insisting 'the areas requiring cleanup are a small portion of the island and Vieques remains one of the few unspoiled, undeveloped areas in the Caribbean …' (Clark, Hagee, and Penn 2005: 13).

Before removing any contamination, the Navy must first get rid of unexploded missiles, rockets, and ordnance. The Navy insists that removal of these materials is too difficult and too costly and says there is no evidence suggesting detonation would pose a risk to anyone (Lindsay-Poland 2006). They also refuse to use special detonation chambers such as those being employed in Massachusetts and other states (Nilda Medina, personal communication, 2006). Instead they began detonation on site in 2005, with the expectation that it will take eight years to finish (Quintanilla 2006: 3).

Within the narrative of the Navy, the value of Vieques was its geographic location. The site was of little concern; the people, their livelihood, and the environment were erased. Vieques embodied these power relations via the physical naval facilities that dominated the landscape and the bombs that fell, and it continues to embody them in the form of contamination, limited or no access to land, high cancer rates, and extreme poverty.

Activists produce scale

Towers (2000: 23) says, 'because of the nature of grassroots resistance, it is limited within its spatial boundaries; thus, it has to strategically construct and make connections to other geographic scales or scales of meaning in order to correspond and effect the larger environmental movement's goals'. CPRDV recognized from the beginning the political need to jump scale. They reached out to Puerto Rican professionals and US and international institutions for help in developing a plan for sustainable development, knowing these associations would make visible the plight of Vieques and give powerful people a personal interest in its future (Robert Rabin, personal communication, 2004). The resulting plan focused on giving Viequenses control over their land and their livelihoods by generating locally-controlled agriculture and small-scale ecotourism.

Zimmerer and Bassett (2003: 6) note that communities are 'characterized as much by their heterogeneity as by their uneasy alliance around certain issues'. As with the rest of Puerto Rico, the question of the political relationship with the US pervaded the politics of Vieques.[2] Many Vieques residents opposed the Navy but did not necessarily support independent status. During the late 1970s when the fishermen protested, they had strategically defined their issues in terms of livelihood and health concerns rather than in terms of freedom from colonialism. CPRDV members, many of whom were involved in the Fishermen's War, appropriated this non-partisan narrative (McCaffrey 2002: 83–87). In addition to the divisions concerning status, there was an obvious split on Vieques between anti-Navy and pro-Navy residents. Most Viequenses and some North Americans wanted the Navy to transfer the land over to the municipality of Vieques. Others wanted the government to retain charge of the lands for one or more reasons: they believed the operations were critical to national security; they believed the Navy was necessary for preserving the vast areas of undeveloped land and maintaining the current low level of tourism; or they believed that the Viequense government would be incompetent if the land reverted to them (Nilda Medina, personal communication, 2002; Levin 2003).

David Sanes' tragic death in 1999 provided what Rodriguez refers to as a 'historical window of opportunity,' in which 'a diverse set of factors converge to generate the appropriate conditions that lead to social action or social movements' (2004: 395). CPRDV understood and responded immediately by organizing the

---

2   There are three extremely oppositional political parties in Puerto Rico: the pro-commonwealth party *Partido Popular Democratico*, the pro-independence party *Partido Independentista Puertorriqueno*, and the pro-statehood party *Partido Nuevo Progresista*. Puerto Ricans have voted on their status in three plebiscites: commonwealth status was strongly affirmed in 1967; commonwealth status won over statehood by two per cent in 1993; 46 per cent chose statehood in 1997, but because 50 per cent chose 'none of the above', commonwealth status remained (Malavet 2004). Another plebiscite is to be held no later than 30 September 2009.

occupation of Camp Garcia. Within hours they had generated media coverage, with dramatic footage of fences being torn down and people rushing into the camp. For four years they worked at jumping scale. They used the site of Vieques as a means of discourse, encouraging local artists to paint murals and signs throughout the island. They built the Camp for Peace and Justice directly across from the gates of Camp Garcia, where volunteers were available around the clock to talk with tourists and visitors (and geographers), and to provide reading materials and plenty of good food. They held regular events, including concerts, political rallies, and discussion groups.

In October 1999, six months after Sanes' death, a Special Panel for the US Department of Defense published a recommendation for 'a working group ... to identify and involve groups linked to the cause of Vieques, especially in Puerto Rican and Hispanic communities, and in environmental and civil and human rights protection groups.' CPRDV, however, had not needed this government report to tell them to form these international coalitions; by the time the report came out, they already had met several times with the Congressional Hispanic Caucus, organized a protest of 50,000 people, made trips to the Puerto Rican and US mainlands, lobbied Congress in Washington, established two bi-lingual Spanish-English websites, and saturated the media, particularly Spanish-language newspapers and television networks.

CPRDV created political alliances at the national Puerto Rican scale by organizing *Todo Puerto Rico con Vieques* (All Puerto Rico with Vieques) (TPRV), made up of community, political, and religious groups, labour unions, NGOs, and schools and universities on the Puerto Rican and US mainlands. TPRV maintained the non-partisan narrative of CPRDV. They worked on establishing reciprocal productions of scale, organizing people from Vieques to travel to the mainlands, and vice versa, giving presentations and meeting with other activist groups.

Puerto Ricans were quick to support the protests, including former Governor Pedro Rossello and Governor Sila Calderon, former Senator Ruben Berríos, the heads of all three political parties, and community leaders. TPRV reached out to celebrities, including singers Marc Anthony and José Feliciano; actors Benecio Del Toro, Andy Garcia, Rosie Perez, and Martin Sheen; and athletes Roberto Alomar, Carlos Delgado, and Chi-Chi Rodriguez. On 16 April, 2001 several celebrities took out a full page ad in the New York Times entitled, 'We ask you to Stop Bombing Vieques Now.' Internationally, the Italian-based Ricky Martin Fan Club, working with TPRV, was instrumental in generating attention.

The three Puerto Rican members of the US Congress, Nydia Velázquez and Jose Serrano (New York), and Luis Gutiérrez (Illinois), were arrested during protests in Vieques and in Washington. Serrano, who carries considerable clout as the Democrat on the House Appropriations Subcommittee, was a key player.[3] So too was Dennis Rivera, president of New York City's Local 1199, the powerful health care workers union. He single-handedly mobilized former presidents Bill Clinton and George H. Bush, New York Governor George Pataki, Robert F. Kennedy Jr., the Reverend Jesse

---

3   In 2004, Serrano secured $1 million in federal funding to help with the cleanup of Vieques (Egbert 2004).

Jackson, and members of Congress who were influential with the Bush White House (Robert Rabin, personal communication, 2004).

In another crucial manoeuvre, CPRDV connected with transnational activists who 'involve themselves in various campaigns out of a moral sense of right and wrong and are not primarily moved by national, cultural or ethnic identities' (Hestres 2004: 60). Some, such as the Dalai Lama, are famous. Most are activists in religious, environmental, antimilitary, civil rights, and human rights organizations including the UN, who organized demonstrations around the world to bring the circumstance of Vieques to international standing.

Observers consistently argue that a key to the success of the Vieques movement was the activists' ability to form an alliance of the US Hispanic community (see, for instance, Masri, 2005; Torres 2005). Hispanic organizations exist, including the Congressional Hispanic Caucus, the US Hispanic Chamber of Commerce, and the National Hispanic Policy Forum. Historically, however, there are tensions among the different ethnic groups and nationalities, especially toward Puerto Ricans, who often are resented for their automatic citizenship. CPRDV's narrative with the Hispanic community emphasized that the problems in Vieques were not just those of the residents; rather, the actions of the Navy violated basic human rights to health and a sustainable environment (Torres 2005: 10). They stressed the global diffusion of environmental pollutants, and denounced the pollution of US military bases around the world. They spoke with Hispanic veterans groups about the health damages from the weapons that had been tested on Vieques, including Agent Orange (Masri 2005). They talked of trans-cultural interests and conditions in the US for Hispanics and people of colour.

CPRDV recognized the importance of reaching out beyond the local Vieques community: when asked why he thought the Navy finally left Vieques, Rabin answered:

> one, it was costing approximately $11 million a year; two, the unrelenting campaign of non violent civil disobedience brought world attention to this issue ... and negative world opinion toward US military, and three, we worked systematically to build strategic relations with economically and political significant Puerto Ricans and other Hispanic groups in the US to get the Vieques issue on the electoral table ... as an issue important to the millions of Latino voters in the US among other reasons (quoted in Small 2005).

Conclusion: After 1 May, 2003

Massey suggests that, 'Places are ... the moments through which the global is constituted, invented, coordinated, produced. They are "agents" in globalization' (2004: 11). This was the case for Vieques. The Navy conceptualized Vieques through a discursive regime in which national security superseded individual rights and the environment, and erased the site of Vieques. It was within this site, however, that new scales of power were produced through a discourse that framed the Navy bombing in Vieques as a violation of *human* rights, effectively collapsing hierarchies and divisions into one scale ('we are the world'). Rodriquez argues that the Vieques protest movement was 'the most significant, comprehensive, and organized social movement ... perhaps in Puerto Rican history' (2004: 396). Robert Rabin, head of CPRDV says, 'We learned that the federal system – when confronted by a determined community – can be defeated' (quoted in Masri 2005: 1347).

Nonetheless, hegemonic power structures are difficult to challenge. As Rodriquez also notes, the success of the movement was contingent upon a window of opportunity: Sanes' death. After 9/11, the anti-Navy protests slowed down. If the US government had changed their mind about closing the base,[4] given the early support for the war in Iraq it is debatable whether CPRDV could have regenerated the momentum – a reminder that the politics of scale is always situated within what Massey calls 'power geometries' (1993). Moreover, the Navy redistributed naval operations from Vieques to various sites in Florida and North Carolina, giving an ironic and disturbing twist to CPRDV's description of their struggle as 'intertwined with the fate of others around the globe bearing the brunt of United States military imperial power' (Grady Flores 2004).

Despite the success of the protests, the future of Vieques is uncertain: will it be a contaminated island with extreme poverty, a low quality of life, and a high death rate? The US Environmental Protection Agency (2005) is no longer predicting how long it will take for cleanup of the contamination. Viequenses still do not have access to the land appropriated by the Navy, the potential for agriculture is grim, there is growing concern about the toxins in the fish catch, and the cancer rate continues to grow.

Will external forces continue to dominate the economic and environmental structures in Vieques? In the fight against the naval base, CPRDV was able to use a rhetoric that effectively countered the national and global needs touted by the Navy. It is harder for CPRDV to argue against the larger-scale benefits of the National Wildlife Refuge, which is a habitat for 114 bird species, tree amphibians, and 16 land reptile species, and protects 15 species of marine and land turtles that were at risk of becoming extinct. Says Refuge Manager Oscar Díaz: 'This refuge has a dry forest. That's a treasure that must be preserved because 94 per cent of all dry forest in Puerto Rico has been destroyed' (quoted in Ruiz 2003), and 95 per cent of what little tourism currently exists is controlled by North Americans (Schmelzkopf, forthcoming).

Will Vieques become another Caribbean tourist trap? In 2003 Damaso Serrano, the mayor of Vieques, wrote a letter to the editor of the local paper, quoting the poet Virgilio Davila and pleading: 'Don't sell your land to the foreigner ... whoever sells his home is selling his homeland' (*Vieques Times*, 5). In spite of his message, many Viequenses *have* sold their homes (without title to the land) as prices have quadrupled. Further, the 42-acre Martineau Bay, Vieques' first $400 a night mega-resort opened in 2006, and a recent article in the New York Times noted: 'a tourism site hot spot (has been) born ... the days of feeling alone are gone ... the crowd has

---

4   Most commentators argue that President Bush kept his word to close the naval base only to avoid alienating the Hispanic coalition prior to the 2004 presidential election (Hestres 2004).

changed ... sun bathers in their 30s shared the beach with rowdy college students playing touch football and drinking beer' (Higgins 2006: 2).[5]

Or will Vieques become a clean, economically stable island with small-scale eco-tourism and agriculture? This is the goal of CPRDV. They have now developed a narrative to renegotiate a *localized* Vieques cut off from the impositions of a broader scale, calling for 'a free Vieques – free from the military, free from the United States Fish and Wildlife Service, and free from land speculators and greedy developers' (Brassell 2005). They want a healthy environment and a good livelihood for Viequenses. Yet they face a difficult challenge. Member Judith Conde says, 'Unlike when we were struggling against the Navy, now we are struggling against something that has no face' (Duff 2004). In additional, Puerto Rican filmmaker Frances Negron-Muntaner observes that once the Navy was gone, 'Vieques as televised melodrama came to an end' (quoted in Gonzalez 2004: 1347).

The life and livelihood of the Viequenses is still threatened by environmental contamination and limited access to the land. Added to that is the very real threat of gentrification from eco-tourism and foreign investment. Yet CPRDV continues to fight for a clean environment, and for Viequenses' right to the island and a high quality of life. Viequenses are left with a qualified victory. It remains to be seen whether they can ever be the ones to decide their own conditions of truth and have the right to build their own livelihoods and landscapes.

Acknowledgements

Many thanks to the people of Vieques who took time to speak with me, feed me, and answer my questions. Thanks to the editors of this book for their insightful suggestions, and thanks to spouse and spawn (Dan and Laura), my ever-patient travelling companions.

References

Agency for Toxic Substances and Disease Registry (2003), 'Final Health Assessment on Vieques, Puerto Rico Finding is No Apparent Public Health Hazard', 15 September, <http:// www.atsdr.cdc.gov/NEWS/viequespr091503.html>.

Baver, S. (2006), 'Peace is More than the End of Bombing: The Second Stage of the Vieques Struggle', *Latin American Perspectives* 33, 102–113.

Berman Santana, D. (2002), 'Resisting Toxic Militarism: Vieques versus the US Navy', *Social Justice* 29, 37–47.

Bird, J., Curtis, B., Putnam, T., Robertson, G., Tickner, L. (eds) (1993), *Mapping the Futures: Local Cultures Global Change* (New York: Routledge).

Brassell, B. (2005), 'Special Committee on Decolonization', United Nations Press Release GA/COL/3121, 1 June, <http://www.un.org/news/Press/docs/2005/gac ol3121.doc.htm>.

---

5 The hotel has changed hands numerous times since construction started in 1999, most recently in 2007 by Wyndham Corp to W Hotels, which will reopen in December 2008.

Clark, V., Hagee, M, and Penn, B.J. (2005), 'Joint Statement before the Subcommittee on Military Quality of Life and Veterans Affairs of the House Appropriations Committee, 9 March, <http://www.navy.mil/navydata/testimony/qol/penn050 309.pdf>.

CNN (2001), 'RFK Jr. Speaks Out Against Bombing on Vieques', 3 August <http://archives.cnn.com/2001/US/08/03/robert.kennedy.jr.cnna/index.html>.

Condit, C. (1987), 'Democracy and Civil Rights: The Universalizing Influence of Public Argumentation', *Communication Monographs* 54, 1–18.

Cox, K.R. (ed.) (1997), *Spaces of Globalization: Reasserting the Power of the Local* (New York: Guildford Press).

Duany, J. (2002), *The Puerto Rican Nation on the Move: Identities on the Island and in the United States* (Chapel Hill: University of North Carolina Press).

Duff, K.G. (2004), 'Buyers Besiege Vieques, Viequenses Losing Their Island?' *Claridad*, 8–14 July, accessed at <http://kbg.web.prdigital.com/ventasesperanza englishwhole.htm>.

Egbert, B. (2004), 'Serrano Plays Santa with 18M', *New York Daily News*, 1 December, S2.

Frommer's (no date), 'Introduction to Vieques' [website], <http://www.frommers. com/destinations/vieques/>.

Goldberg, M. (2004), 'Primeval Minefield', *The American Prospect* 15: 11, 7.

Gonzalez, R. (2004), 'Boricua Gazing: An Interview with Frances Negron-Muntaner, *Signs* 30, 1345–1351.

Grady Flores, M.A. (2005), 'Special Committee on Decolonization', United Nations Press Release GA/COL/3121, 1 June <www.un.org/news/Press/docs/2005/gacol3 121.doc.htm>.

Grusky, S. (1992), 'The Navy as Social Provider in Vieques', Puerto Rico, *Armed Forces and Society* 18, 215–230.

Herod, A. and Wright, M. (2002), 'Introduction: Rhetorics of Space', in Herod and Wright (eds).

Herod, A. and Wright, M. (eds) (2002), *Geographies of Power: Placing Scale* (Malden, MA: Blackwell).

Hestres, L. (2004), 'Peace for Vieques: The Role of Transnational Activist Networks in International Negotiations', MA Thesis, Georgetown University.

Higgins, M. (2006), 'Is Success Spoiling Vieques?' *New York Times*, 2 April, 5, 2.

Johnson, J.L. and Jones, J. (1999), 'Vieques Must Not Close; Getting Beyond the Training Incident', *Washington Times*, 30 September, A21.

Levin, S (2003), 'Springtime for Vieques … 2003', <http://www.enchanted-isle. com/enchanted/letters/lvspring03.htm>.

Lindsay-Poland, J. (2006), 'The Future of a Bombing Range', *Peace and Change* 31, 136–8.

Lloréns, H. (2006), 'Dislocated Geographies: A Story of Border Crossings', *Small Axe* 19, 74–93.

Malavet, P.A. (2004), *America's Colony: The Political and Cultural Conflict between the United States and Puerto Rico* (New York: New York University Press).

Masri, R (2005), 'The Price of War Games', *Southern Exposure* 32, Winter, <http://www.southernstudies.org/reports/Masri3-WEB.htm>.

Massey, D. (1993), 'Power-Geometry and a Progressive Sense of Place', in Bird, et. al. (eds).

Massey, D. (2004), 'Geographies of Responsibility', *Geografiska Annaler B* 86, 5–18.

McCaffrey, K.T. Military (2002), *Power and Popular Protest: The US Navy in Vieques, Puerto Rico* (New Brunswick, NJ: Rutgers University Press).

McCaffrey, K.T. (2006), 'The Battle for Vieques' Future', *Centro Journal* 18, 125–145.

Minca, C. (2001), 'Postmodern Temptations', in Minca (ed.).

Minca C. (ed.) (2001), *Postmodern Geography: Theory and Praxis* (Malden MA: Blackwell).

Navy Environmental News (1999), 'RADM Granuzzo Testifies to Rush Panel', #99–06 NEN99041, 16 August.

Ortiz-Roque, C. and López-Rivera, Y. (2004), 'Mercury Contamination in Reproductive Age Women in a Caribbean Island: Vieques', *Journal of Epidemiology and Community Health* 58, 756–7.

Purcell, M. and Brown, J.C. (2005), 'There's Nothing Inherent about Scale: Political Ecology, the Local Trap, and the Politics of Development in the Brazilian Amazon', *Geoforum* 36, 607–624.

Quintanilla, R. (2006), 'In Vieques, Bombing Resumes Ordnance Detonation Postpones Start Of Environmental Cleanup', *Chicago Tribune*, 27 March, 3.

Rodríguez, H. (2004), 'A "Long Walk to Freedom" and Democracy: Human Rights, Globalization, and Social Injustice', *Social Forces* 83, 391–413.

Ruiz, C. (2003), 'Who Will Clean Up the Navy's Toxic Mess in Puerto Rico?' *Inter Press Service* (IPS), 31 December, < http://us.oneworld.net/article/view/76053/1/>.

Ryder, K.M. (2004), 'Comment: Vieques' Struggle for Freedom: Environmental Litigation, Civil Disobedience, and Political Marketing Proves Successful', *Penn State Environmental Law Review* 12, 419–44.

Schmelzkopf, K. (forthcoming), 'Neocolonialism and the Control of Tourism by North Americans in Vieques, Puerto Rico', *Tourism Geographies*, under review.

Small, M. (2005), 'Now We Are Free – Lessons from Vieques', Indymedia UK, <http://www.indymedia.org.uk/en/2005/04/309142.html>.

Stott, P. and Sullivan, S. (2000), 'Getting the Science Right, or Introducing Science in the First Place?' in Stott and Sullivan (eds).

Stott, P. and Sullivan, S. (eds) (2000), *Political Ecology: Science, Myth and Power* (London: Edward Arnold).

Swyngedouw, E. (1997), 'Neither Global nor Local: 'Glocalization' and the Politics of Scale', in Cox (ed.).

Torres, M.I. (2005), 'Organizing, Educating, and Advocating for Health and Human Rights in Vieques, Puerto Rico', *American Journal of Public Health* 95, 9–12.

Towers, G. (2000), 'Applying the Political Geography of Scale: Grassroots Environmental Justice', *Professional Geographer* 52, 23–36.

US Armed Services Committee (1999), *Vieques and the Future of the Atlantic Fleet Weapons Training Facility* (Washington, DC: US GPO).

US Department of Defense (1999), 'Report to the Secretary of Defense of the Special Panel on Military Operations on Vieques', <http://www.defenselink.mil/news/Oct1999/viq_101899.html>.

US Environmental Protection Agency (2004), 'EPA Proposes the Atlantic Fleet weapons Training Area on Vieques and Culebra for Inclusion on the Superfund National Priorities List', <http://www.epa.gov/superfund/news/npl_081304.htm>.

Vieques Times (2003), 'Letter to the Editor', Winter, 5.

Welcome to Puerto Rico (n.d.), 'Vieques', <http://welcome.topuertorico.org/city/vieques.shtml>.

Wilcox, J. (2001), 'Vieques, Puerto Rico: An Island under Siege, *American Journal of Public Health* 95, 695–701.

Zimmerer, K.S. and Bassett, T. (2003), 'Approaching Political Ecology: Society, Nature, and Scale in Human-Environment Studies', in Zimmerer and Bassett (eds).

Zimmerer, K.S. and Bassett, T. (eds) 2003, *Political Ecology: An Integrative Approach to Geography and Environmental Development Studies* (NY: Guilford Press).

Chapter 9

Making Local Places GE-Free in California's Contentious Geographies of Genetic Pollution and Coexistence

Dustin Mulvaney

"Shall the people of Mendocino County prohibit the propagation, cultivation, raising and growing of genetically modified organisms in Mendocino County?" – Anti-GE activist (quoted in Hamburg 2003).

"I don't care what goes on in Mendocino ... They can get high on marijuana, sit around eating organic food and all be thinking that somehow they're living healthier lives than the rest of us. I think it's a joke" – Biotechnology Industry Organization (Somers 2004).

Introduction

What does it mean to be in a GE-free zone and how are those places made? When anti-genetic engineering activists in Mendocino County helped pass a county ban on genetically engineered organisms (GEO), questions about its significance immediately surfaced. Activists were quick to claim a new era for the regulation of GEOs. On the other hand, the Mendocino GE-free zone seemed irrelevant to GEO regulation as none are grown there. Yet the GE-free zone was vigorously opposed by industry. The ban set a worrying precedent for the regulation of agricultural biotechnology. Other California counties soon followed suit with similar GE-free zones. By 2006 four counties claimed to be GE-free zones and 13 other counties considered similar bans (Map 9.1). Meanwhile, 12 counties in California passed resolutions declaring a commitment to the *use* of GEOs. This fracture between counties claiming to be GE-free zones and counties promoting GEOs maps onto California's contentious geography of industrial and alternative agriculture.

A central contention in debates about GEOs is their containment. Scientists describe a multitude of potential ecological risks from GEOs, but the risks come with many uncertainties. Should planting GEOs be prohibited until the uncertainties are better known? Or, is it possible to contain the genes of GEOs and preserve the identity of food from field to plate?

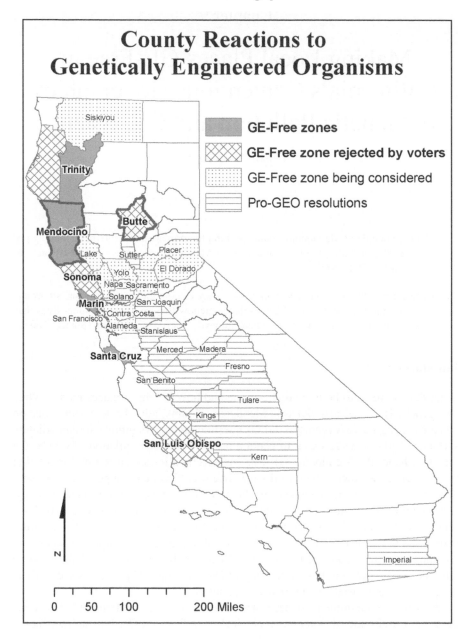

Map 9.1   County actions on genetically engineered organisms in California

The purpose of GE-free zones is to keep out *genetic pollution* – the inadvertent contamination of food and seed with unwanted genes.[1] Several notable incidents where GEOs emerged in places they did not belong helped activists demonstrate the challenges associated with containing living organisms (see below). Anti-genetic engineering (anti-GE) activists forcefully maintain that the complexities of managing gene flow and identity preservation make genetic pollution inevitable. Recalling a gene from an organism, or an organism from the environment, is difficult if not impossible. The potential for significant and irreversible consequences leads activists and ecologists to suggest a precautionary approach to GEOs.

The very notion of genetic pollution is contested by the agricultural biotechnology industry. Genetic pollution is in their view adventitious presence – incidental living materials of little to no significance. Recognizing that adventitious presence poses some market consequences, GEO proponents emphasize *coexistence*. The concept of coexistence requires *identity preservation* practices that allow GEOs to be grown alongside conventional and organic crops, without affecting their 'marketability'. Identity preservation keeps GEOs in fields spatially distant from other crops and adopts handling practices designed to prevent commingling, such as dedicating equipment to GEO fields and ensuring seed and product is properly labelled. Coexistence attempts to keep GEOs in the fields they are grown. This distinction between keeping GEOs *in* (contained) versus keeping them *out* (banned) is central to California's contentious geography of genetic pollution and coexistence.

This chapter examines two case studies of GE-free zone ordinances in California. The first is a contest over an ordinance in Mendocino County, the first US county-level GEO ban. In the second, Butte County, a similar ordinance failed. Activists view local GE-free zones as a response to gaps in the national regulatory system as well as a symbolic response to the cession of biosafety authority to international trade tribunals (Henson, 2001). Responding to these gaps and challenges to governance, anti-GE activists are exerting territorial authority in new arenas.

The uses of space, appropriate scale of regulation, and particular place-based consequences are center stage in GEO controversies. Different interests assert control over spaces and places, often by rescaling spatial representation or organization. Rescaling implies a re-ordering of space: global, national, regional, local, community, household, body, genome, and gene. The scalar order is socially produced, not ontologically preordained by biophysical features (Smith 1992). The politics of scale reveals power asymmetries between different actors: some can exercise power in local government, in households, or over their bodies, while others exercise power at the state, national, or even global scales (Swyngedouw 1997).

From body politics to global trade politics, scalar processes shape where GEOs are grown and eaten. Anti-GE activists and seed firms strategically 'jump scale' to arenas more favorable to their interests and politics (Smith 1992). Multinational seed corporations and their proxies mobilize legal firms and lobbying efforts at global and national scales, where science and intellectual property rights dominate discourse

---

1 The term biological pollution – organisms in places they do not belong – is also relevant to the ecological risks of GEOs. For the sake of clarity here, genetic pollution is taken to be synonymous with biological pollution.

and practice. Anti-GE activists work to enroll local jurisdictions and consumer perceptions by appealing to local narratives and bodily interests. Activists engage at scales that are effective as consumers and local citizens. In doing so, they attempt to by-pass federal authority, tempting fate with legal precedence.

Yet, there is limited explanatory power in how social groups exercise power through an analysis of the politics of scale alone. The politics of *place* can aid explanations for why similar processes have different outcomes, such as when Canada decided to ban the use of Monsanto's recombinant-bovine growth hormone (rBGH) and the US decided to approve it for wide-spread use (Mills 2002). Each place is constituted by a different constellation of social experiences, values, and knowledge, with different relationships to nature and culture, producing different political persuasions and ideas about environmental governance.

Place-based differences also highlight how apparently similar discourses – assemblages of images, concepts, categories, interpretations, and contexts embedded in language and shaping social practice (Hay 2000) – take on different meanings. Groups reach shared understandings of the world through discourses that give meaning to experiences. Discourse coalitions are groups with collective understandings of phenomenon, which share common sets of storylines, which are the discursive elements of larger narratives that give meaning to phenomenon and provide the basis for social action (Hajer 1995). A particular storyline identifies a problem, assigns it a causal lineage, and asserts a remedying action. Opposition to unintended GEOs in the food system constitutes a discourse coalition, where the storylines of adventitious presence and genetic pollution overlap in achieving similar aims: keeping GEOs contained. Given that the storylines to seek GEO containment are very different, coalitions can enlist broader support for social action, but coalitions can also be contradictory and short lived.

The politics of scale, place, and discourse help explain social movement efficacy. Social movement theory unpacks how internal organization, resources, external factors, and discursive framing shape social movement success. Emerging out of sociology and political science, most of this work lacks a spatial component (Miller 2000). The spatial element to social movements is central to ideas like *militant particularism* (Williams 1989) – which involves a characterization of the local – *network society* (Castells 1996) – which incorporates changing spatial relations – and *political ecology* (Peet and Watts 2004) which looks at how local and global forces interact. These ideas are essential to any engagement with social change, because power asymmetries shape the contours of social agency. Examining power asymmetries can reveal the place-based and scalar components of social movement efficacy. Some strategies may be more effective at some scales and in some places, but not others. This is evident with anti-GE activists whose discourse is credible and appealing in one place, but not another.

Anti-GE Activist Storylines

Anti-GE activists opposing GEOs in agriculture represent a diverse set of interests. Motivations stem from concerns about food safety (Nestle 2003), ecological risk

(Snow et al. 2004), corporate control of the food system and technoscientific research (Heffernan 1999; Kloppenburg 2004), property rights in living organisms (Rifkin 1998), and university privatization (Kenney 1986). Acting at different scales and places, and navigating multiple discursive terrains, anti-GE activists hope to influence the social and environmental dimensions of GEOs, redistribute the risks and benefits, and ultimately redirect the trajectory of the agro-food system.

The *ecological risks* of GEOs are broadly described in the scientific literature (Letourneau and Burrows 2002; Rissler and Mellon 1996; National Research Council 2002). Potential impacts include producing GEOs that behave like invasive species or that will hybridize with sexually compatible species. Hybridization and subsequent introgression can foster biodiversity loss through genetic swamping or inbreeding depression, particularly in centers of crop genetic diversity (Fowler and Mooney 1990; Ellstrand et al. 1999) or in areas where wild species are endangered (Kelso 2004). Harms to non-target organisms from toxic pollen are another ecological risk, as illustrated after scientists showed the increased mortality of monarch butterfly larvae eating milkweed dusted with Bt *(bacillus thuringiensis)* pollen (Losey 1999). The considerable challenges in recalling living organisms from the environment, or genes from genomes, have led many ecologists to call for stricter evaluation and regulation of GEOs because of potentially irreversible ecosystem harms (Snow et al. 2004; Kapuscinski and Hallerman 1990). Ecologists suggest that current approaches to risk assessment need more temporal and spatial breadth, and site specificity, before making broad claims about biosafety (Kapuscinski 2002; National Research Council 2002).

Food safety advocates argue that little is known about the long-term health consequences of GE foods (Nestle 2003). Potential risks include new food allergies, anti-biotic resistance, or foods with increased toxicity. The use of food crops to produce pharmaceuticals, biologics, and industrial compounds also pose serious food safety consequences (Union of Concerned Scientists 2003). Food safety regulation is based on the controversial principle of "substantial equivalence," which compares the characteristics of GE foods (e.g. protein, carbohydrates, fatty content) with their conventional counterparts and suggests that if they are similar, they should be regulated similarly. This assessment of food safety is contested as being overtly reductionist and is criticized for failing to account for, among other things, the use of viral promoters, which stay in the organism.

The introgression of herbicide tolerance traits into weeds or increased quantities of herbicide could produce *herbicide resistant weeds* (Gould 1998). Herbicide resistant weeds could provoke the use of more toxic chemicals (Benbrook 2003). The widespread adoption of Bt crops that express endotoxins in plant tissues to kill *Lepidopteran* and *Coleopteran* insects raises concerns about the evolution of *insect resistance* (Alstad and Andow 1995), which would undermine the effectiveness of a topical, non-toxic form of Bt crystals used by organic farmers (McGaughey et al. 1998).

Growers face the prospect of lawsuits for *patent violations*. Monsanto has pursued over 147 growers to secure seed licensing fees (Center for Food Safety 2005). Percy Schmeiser, a Canadian canola farmer, was sued after Monsanto's RoundUp Ready canola was found in the fields of his family farm. Even though the crop was not intentionally planted and possibly entered the field from passing trucks that transport canola on adjacent roads, a subsequent court ruling explicitly maintained that its presence alone constituted a patent infringement. Under the current interpretation of law, growers are responsible for keeping patented pollen and seed out of their fields. If a seed contaminates a neighbour's crop, affecting the grower's seed stock or market price, no firm or grower is liable.

Additionally, growers are increasingly aware of *market risks*. Some nations have particularly strict biosafety requirements for imports. When specified safety requirements are not met, a country may shut down trade immediately. For example, Japan banned US beef after flesh from a cow's spinal chord was found in beef imports. The nervous system tissues of cows are believed to harbour mad-cow disease and Japan requires the tissues be removed before import. Markets may even shut down the suspicion of being unsafe, as when 58 countries banned US beef imports after single US Holstein cow tested positive for mad cow disease.

Similar but more formal biosafety regulatory regimes exist for GEOs. As permitted under the Cartagena Protocol on Biosafety, many nations prohibit GEOs, set tolerance thresholds, or require labelling. Failure to comply with these requirements, has led to the loss of export markets for some growers. In 2006, an unapproved variety of GE rice known as LibertyLink® was found in long grain rice from the US South. This led to immediate bans by Japan and the European Union on all long grain rice imported from the US.

Activists have contested the *commodification of life* since the Supreme Court ruled in *Diamond v. Chakrabarty* (1980) that living organisms and their parts could be patented. Activists assert that life's parts and processes are products, not of human invention, but of nature and a common human heritage. Even where human coevolution with domesticated plants demonstrates the role of human ingenuity, the fact that only modern plant breeding techniques are patentable undervalues the labor of centuries of farmer selection.

Also, anti-GE activists relate the consequences of the *gene revolution* to the negative implications of the Green Revolution, which led to economic concentration and monopolistic tendencies (Heffernan 1999; Boyd 2003). The advent of high yielding crop varieties might further marshal in an era of even more overproduction, greater rural inequality, and hasten the migration of the rural poor into urban centers. According to activists, GE crops will reinforce these trends. Since the same companies that sold environmentally harmful agrochemicals during and after the Green Revolution are producing GEOs, claims about the environmental benefits of GEOs are treated with suspicion.

Finally, some activists argue that GEOs are *unnatural*, asserting that genetic engineering is radically different because it can breach the species barrier (Howard and Rifkin 1977). Some techniques use viruses and bacteria to invade the nucleus of cells, a process not often found in nature. This is the basis for much of the religious opposition, which contend that crossing species' barriers amounts to 'playing God'.

GE-free Zones and Rescaling for Local Regulation

Activists' formal access to national regulatory discussions is limited to the legal efforts of public interest groups. Access to international dialogues is even more restrictive. Activists are overmatched in lobbying efforts, while government officials work with industry to promote national competitiveness and regional economic growth (Wright 1994; Gottweis 1998). Activists' response is to rescale GEO regulation to the local level by creating GE-free zones. In their view, federal oversight is ineffective. As one activist described, 'we wouldn't have to do this if government was doing their job'. Local government is an appealing venue because activists can connect the potential repercussions of GEOs directly to local people and places.

Antecedents to the GE-free zone strategy include local ordinances banning the initial GE experiments in the 1970s (Krimsky 1982) and a temporary ban on the initial outdoor field tests of a GE bacterium called 'ice-minus' during the mid 1980s (Krimsky and Plough 1988). GE-free zones sprung up across Europe where locales in Austria, France, and Italy asserted jurisdictional authority over rules developed by the European Union. In the US, Vermont activists started the first GE-free zone campaign and within three years seventy Vermont towns declared themselves GE-free.

Around the same time, members of the Californians for GE-free Agriculture formed the Biodemocracy Alliance to help turn California counties into GE-free zones and to prevent the agricultural biotechnology and seed industries from developing preemptive legislation. Many of the Alliance's members and constituents are committed to organic and sustainable agriculture. Some work for state and national organizations, while others work at the grassroots level. The basis for their opposition to GEOs is to ensure that non-GE alternatives are preserved. Because avenues for gene flow cannot be 100 per cent contained, they argue that creating GE-free spaces is critical to ensuring that GEOs do not contaminate food and seed.

To underscore the need for GE-free zones, activists highlight cases where regulatory agencies were unable to prevent genetic pollution. In one case, a variety of Bt corn called StarLink™ was found in Taco Bell taco shells on supermarket shelves throughout the US despite not being approved for human consumption (Segarra and Rawson 2001). In another case, ProdiGene's GE corn that produces pig vaccine was mixed with soybeans in Iowa and Nebraska (Fox 2003). In a third case, patented GE corn traits were discovered in corn landraces in Oaxaca, corn's cradle of crop biodiversity (Quist and Chapela 2001). These incidents, along with several others bolstered support for activists' assertion that it is impossible to prevent the unintended escape of GEOs.

The Biodemocracy Alliance views GE-free zones as more than an attempt to create GE-free spaces across the state. While some hope other counties pass similar ordinances, others hope to initiate a 'crisis of jurisdiction', where local authorities challenge the legitimacy of higher scales of government. Crises of jurisdiction often are symbolic, but in some cases they prompt other scales to react through legislation or the courts.² Specifically thought, this strategy requires enrolling *local* support, which, as this case study explores, activists mobilized at the local scale in two different places.

The Contest Over 'Measure H' in Mendocino

Mendocino County became the first county in the US to ban the cultivation, propagation, or raising of GEOs when 59 per cent of voters approved a ballot initiative – Measure H – in March 2004. The ban was declared a watershed moment for anti-GE activists despite questions about its significance. Responding to the ineffectiveness of federal oversight, activists felt they had no other choice but to make counties GE-free. Situated along the coastal range of California's north coast, Mendocino County is renowned for its organic agriculture, organic and biodynamic wines, and counter-culture roots. With over 150 organic farms and wineries, alternative agriculture is part of the county's identity. While the anti-GE storylines above provided the basis for anti-GE activism, a primary discursive tactic employed and emphasized here was how local agriculture needed to be defended from genetic pollution.

Genetic pollution not only poses ecological risks, but also more immediate economic consequences to local growers. One grower claimed that 'contaminated crops may not get the organic premium, or worse, may get us sued for violating a patent.' Genetic pollution could also affect the perceived quality of products from Mendocino. Price premiums and niche markets make Mendocino's smaller-scale production profitable and provide economic incentives to keep lands in agriculture. This discourse enrolled key allies, including the country's largest organic wine grape grower. Many wine grape growers supported Measure H despite claims that the ban would prevent growers from getting new tools to deal with the glassy-winged sharpshooter, a vector for Pierce's Disease, an emerging pest in California grapes. Support for the ban was fuelled by fears that genetic pollution would cause importers to reject their wine. Earlier in 2001, the county's largest wine grape grower announced the active promotion of their wines as GE-free (Merrill and Culp 2001). The company president said that GE wine grapes would counteract a commitment to long-term sustainable agricultural practices: 'We've committed part of our system to organics and GE poses a threat to our market position.' This was echoed throughout the Mendocino wine industry. A wine wholesaler explains, 'organic winemakers and importers ... are very concerned about the contamination of organic vineyards by wind- and insect-born GE pollen. As the proud custodians of the purity of their fruit, organic vignerons have much to fear from GE. If there is accidental contamination the damage will be irreversible' (Merrill and Culp 2001). Growers were well aware of the market implications of genetic pollution. One organic grower said, 'being GE-free has advantages from a marketing perspective.'

---

2   For example, in the late 1970s, Mendocino County barred the California Department of Transportation from applying herbicides on roadsides after a publicized incident where herbicides were found coated on county school buses. The state reacted with preemptive legislation that reorganized the oversight of agrochemicals, a move that made regulation more effective.

Groups like the County Farm Bureau contested the benefits of the GE-free zone to growers, suggesting impacts to overall economic competitiveness. The California Plant Health Association, a chemical industry trade association, said the ban would deny growers a tool they may someday need. One grower emphasized the foreclosure of technological opportunities: 'What happens if Napa and Sonoma grape growers end up being able to use beneficial new grapevines, but we can't?' (Geniella 2004).

Despite debates on the pros and cons of the GE-free zone, the enforcement of the ban was itself a central point of contention. If the ban were to affect growers, it would have to be enforced. The county agricultural commissioner said he had 'no plans to broaden his office's duties to enforce the law' (Lucas 2004). Ban opponents argued that the ban 'would create a new government program that requires additional funding that we don't have,' enrolling support from the anti-tax demographic by implying that a ban required new taxes for monitoring and enforcement.

Ban opponents creatively dealt with the issue of surveillance and monitoring. In a state known for extensive industrial agriculture, California's true top agricultural commodity is marijuana (Gettman and Armentano 1998). As part of the 'green triangle' with Humboldt and Trinity counties, Mendocino is a significant producer. As the vote approached, radio ads asked voters how they felt about backyard inspections (for GEOs), explicitly playing on the fears of those growing illicit crops.

Anti-GE activists, however, created strategic political spaces by raising locally sensitive environmental issues. Activists emphasized the implications of GEOs to fishing communities, gaining support from the Pacific Coast Federation of Fisherman's Associations, the West Coast's largest commercial fishing association. They cited a study showing that herbicide tolerant crops would increase herbicide use and degrade water quality (Benbrook 2003). Mendocino's rivers support Coho salmon and Steelhead, but today their range is considerably less and their populations are in decline. Activists raised concerns about the ecological impacts of GE salmon (Kelso 2004), which had recently made local headlines when the California legislature authorized the Department of Fish and Game to prohibit GE salmon and implement a permitting process for GE fish.

An effective tactic used by activists was to unmask the outside interests driving the opposition to the ban. CropLife America and the California Plant Health Association represented Monsanto, Dow AgroSciences, Bayer CropSciences, Dupont, and other seed and agricultural biotechnology companies. Mendocino County has a long history of outside influence shaping natural resource issues as with offshore oil drilling (Freudenburg and Gramling 1994) and timber harvesting (Mendocino County Historical Society 1996). Opponents of the GE-free zone spent over $600,000 to oppose the campaign, outspending ban proponents almost eight-to-one. This allowed activists to frame the issue as 'Mendocino County versus multinational corporations.' By casting ban opponents as outsiders, activists could question company motives. Concerns about multinational corporations fed into concerns about surveillance of patent infringements. Activists drew on the narrative of Percy Schmeiser and his battle against Monsanto that he recounted on tours of Mendocino as the vote approached.

Since no GE crops were currently grown in Mendocino County, the approval of Measure H was seen as symbolic. Some argued the symbolism was especially important because the county's organic agricultural identity was at stake. For others, the GE-free zone was a response to the increased corporate control of the food system: 'They [Monsanto] can be stopped if we have laws in our counties that forbid the planting of GMOs. We're saying not in our backyard, not in our county' (Hamburg 2003).

In Mendocino, arguments for a GE-free zone ordinance were contested, but influential organic growers lent authenticity to activists' claim to represent local farmers and businesspeople. Ban opponents were challenged to find farmers speaking out against the ban. Yet anti-GE activists – networked across space and scale – drew on outside people and resources as well. What made activists' appeal effective were shared visions and values regarding corporate control, sustainable agriculture, and the environment. Activist rhetoric resonated with many voters, sharing a common vision of the future of Mendocino County and the food system. Anti-GE activists encountered favorable power asymmetries at Mendocino's local scale, where voters and activists share similar interpretations of the solution to genetic pollution.

The Contest Over 'Measure D' in Butte

In November 2004, 61 per cent of Butte County voters rejected Measure D, a ballot initiative similar to the Mendocino GE-free zone. Butte County extends from the western foothills of the Sierra Nevada to the flat expanse of California's Sacramento Valley. Rice is Butte County's number one crop, adding more than $112 million in revenues to the county economy in 2003. California's rice production is energy intensive and highly mechanized. Aerial seeding and herbicide application, and mechanical harvesting mean that human hands rarely touch California rice. Intensive production relies on new technologies, particularly new varieties of rice. Thus a key issue raised during the ballot showdown was how a ban in Butte County would affect the adoption of GE rice in the heart of California's rice country.

California's rice industry relies on Pacific Rim export markets for sales of medium-grain *japonica* rice. As much as 40 per cent of the rice crop is exported to Japan alone. These markets have strict biosafety requirements for GEOs. The rice industry is concerned that genetic contamination would negatively affect access to those markets, which are described as 'thin, volatile, and risky' (Economic Research Service 2001). The volatility of these markets is partly attributable to a WTO requirement that Japan liberalize its rice sector. Today Japan imports rice from California at prices that significantly undercut the prices of Japan's domestic rice producers. Fearing market disruption, in 2000, the California Rice Commission aligned with anti-GE activists in support of the California Rice Certification Act. The law allows the industry to convene an advisory panel to evaluate the introduction of new rice varieties with traits that might impact the marketability of rice. The advisory panel recommends practices to contain gene flow and ensure identity preservation to keep GE and non-GE separated. This helps growers prevent 'genetic pollution' and maintain market access.

Of all California counties with significant industrial agricultural production, Butte County would be the most likely to adopt a ban, given the presence of market concerns raised by the rice industry, some significant players in organic agriculture, and a student demographic. Yet the coalition of anti-GE activists and rice growers that supported the Rice Certification Act did not develop in support of the Butte County GE-free zone. The rice industry maintained that the GE-free zone would compromise its autonomy to make decisions about practices and technologies. GE rice could be a tool to replace more toxic herbicides being phased out by California regulators. In light of the Rice Certification Act, the ban was framed as being redundant. Hence, the Rice Commission voted 28–1 to publicly oppose the GE-free zone.

The failure of a coalition to form between agricultural interests and anti-GE activists shaped the in-effectiveness of the anti-GE discourse; as one official put it, 'that nonsense may work in Mendocino County, but this GE-free zone agenda is certainly not in the interests of farmers here in [Butte] county.' Opponents of Measure D – the Citizens for Responsible Agriculture – cast the GE-free zone as driven by outsiders. They publicly suggested that Bay Area liberals were behind the ban: 'Here you have these liberals in San Francisco and [Washington] DC, who know nothing about farming here, telling us what we can and shouldn't use.' Though the GE-free zone had support from the state's largest organic rice producer, opposition from the Butte County Growers Association and the County Farm Bureau undermined the credibility of activists dealing with farm related issues. Butte County is a conventional agricultural landscape, linking industrialized agriculture to the industrial food system. The ideology of sustainable agriculture is not as closely bound to Butte County's agricultural identity. Here organic farming practices are viewed more as a niche than as the future of agriculture.

One reason the initiative failed was tactical. Ban opponents argued that the language of the measure would prevent people from getting access to medicines, such as GE insulin: 'Many people said that the fear of losing access to medical drugs is why they voted "no" in the end' (McCarthy 2006). They also argued that the ban would affect mutagenesis rice, a commonly grown variety bred using radiation: 'We've worked long and hard to develop these varieties, which are the highest yielding in the word. With one stroke, they'll all be lost to the industry.' One grower argued, 'we have the potential of losing our current high yielding varieties … at the farm level, this would take us back 20 years' (Lee 2004).

Anti-GE activists were explicit about research exemptions. One carefully noted that 'we want to be careful not be against scientific research … we're not.' But while activists drafted an exemption for university research, it did not exempt research at the Biggs Rice Experiment Station, which is cooperatively, but privately, owned by the rice industry. Research scientists, farm organizations, and biotechnology industry representatives effectively persuaded the public that, 'the law means … we can't do anything in recombinant DNA research. Period. Not in the lab. Not in the greenhouse. Not in the field. Not at all' (Lee 2004).

Unlike the Mendocino case, the authenticity of the claims made against the GE-free zone was not undermined by the presence of national agricultural biotechnology trade associations, like CropLife America and the Cattlemen's Association. For one, those groups have more credibility here, often directly in contact with farmers through trade shows, product advertising, and regulatory position papers. But also, learning from the Mendocino campaign, these organizations did not fund the campaign directly, instead funding the campaign through the County Farm Bureau (a tactic that was denied).

The Butte County case reveals how an analysis of the politics of scale, place, and discourse can help explain what appear to be contradictions vis-à-vis the Mendocino case: Both the rice industry and activists share an interest in preventing GEOs from unintentionally entering the food system. But genetic pollution – GEOs 'out of place' (Douglas 2002) – assumes different meanings for different groups. The rice industry translates genetic pollution into a temporary problem of market risks. Out of place GEOs put markets at risk given volatile rice trade with the Pacific Rim. If and when the market risks subside, the rice industry wants to keep open the option of using GE rice in the future. Even if market risks remain, the rice industry maintains that GE rice can be contained because rice is a self-pollinating crop and the industry already employs identity preservation practices to keep different rice varieties separate.

For anti-GE activists, genetic pollution was tied to a broader set of storylines, some emphasizing scientific risks and others without a connection to pollution at all. For some anti-GE activists, the notion of genetic pollution is a discursive tactic. One activist revealed that,

> [p]eople just aren't interested in the mundane politics of agricultural research. I am. But most people won't support or even understand that campaign. Say it will be bad for the environment or food. That gets attention.

These different interpretations of genetic pollution meant little in the context of the Rice Certification Act; both sides agreed that unwanted GEOs in the food system were a bad thing. But when activists infringed upon the autonomy of the rice industry, the genetic pollution discourse coalition fell apart. The GE-free zone would have different implications for the rice industry. Both policies attempt to regulate genetic pollution, but one kept the rice industry autonomous from outside interference and was quite narrowly focused on market risks. As activists engaged with the local scale, they lost the support of the rice industry because a GE-free zone would eliminate GE rice from the set of available tools. One grower said, 'We are already facing global economic competition and increased regulatory burdens, this ban could be another nail in the coffin.' Unlike Mendocino County, where agricultural interests defended the GE-free zone on the grounds of market concerns and economic competitiveness, in Butte the GE-free zone was considered redundant. The California Rice Certification Act already provided a measure of protection against genetic pollution and became the dominate storyline that won the 'legislative' and 'coexistance' day.

Conclusion

This study highlights the importance of contentions over meaning in analyses of political difference and environmental problems. Even a seemingly unitary subject like genetic pollution takes on different meanings for people in different places. These meanings are tied to a sense of authenticity, which is often marked by the ways that different actors are cast as insiders or outsiders. In Butte County, genetic pollution was a manageable problem and the means for managing that problem should remain the province of the rice industry. In Mendocino, genetic pollution was inevitable, irreversible, and significant enough to warrant a GE-free zone. Here, activists' narrative cast those that opposed the GE-free zone as the same outsiders that sought to drill for oil off the coast and clear-cut old growth redwoods, appealing to voters' environmental and aesthetic values.

These cases also show the importance of place in understanding and exploring environmental contentions. While the process of scaling plays on favourable power asymmetries, going local does not always yield the same results. Social power is not tied to a particular scale, but is a constellation of discourses, scales, and places networked through space. Analyses can better reveal power asymmetries by incorporating political projects that are simultaneously unfolding across space and the ways those projects touch down in specific places, operate at different scales, and through specific discourses.

Anti-GE activists recognize the shortcomings of a county-by-county approach, pointing out that GEOs, or any living organisms for that matter, do not recognize political boundaries. How far will symbolic approaches take activists? Do GE-free zones help activists shape the social and environmental consequences of GEOs? These remain the critical unknowns. Whatever the material implications of GE-free zones, they have certainly taken hold in the public eye. Activists are pointing out how these local ordinances are part of a wider agricultural sustainability movement engaged in multiple places and at multiple scales. The Biodemocracy Alliance is getting support in the legislature for a liability bill to hold seed firms liable for the impacts of genetic pollution. But some argue that the GE-free zone as an activist tactic must be reexamined. The agricultural biotechnology and seed industries have responded to GE-free zones with state-level legislation intended to preempt county bans – in other words, a banning of GE bans. From 2005–2006 there were several attempts to pass preemptive legislation (Associated Press 2006) that have since become known as the so-called 'Monsanto laws.' Big business too plays on the politics of scale to take advantage of favourable power asymmetries. But just how this maps onto California's contentious geography of genetic pollution and coexistence remains to be seen.

References

Alstad, D. and Andow, D. (1995), 'Managing the Evolution of Insect Resistance to Transgenic Plants', *Science* 268, 1894–1896.

Associated Press (2006), 'Genetic Crop Bill OK'd by California Assembly', *Monterey Herald* (Monterey, California).

Benbrook, C. (2003), 'Impacts of Genetically Engineered Crops on Pesticide Use in the United States: The First Eight Years', Biotech InfoNet.

Boyd, W. (2003), 'Wonderful Potencies? Deep Structure and the Problem of Monopoly in Agricultural Biotechnologies', in Schurman, R. and Takahashi Kelso, D. (eds), *Engineering Trouble: Biotechnology and its Discontents* (Berkeley: University of California Press).

Castells, M. (1996), *The Power of Idenity* (Cambridge: Oxford University Press).

Center for Food Safety (2005), 'Monsanto vs. US Farmers', Washington DC.

Douglas, M. (2002), *Purity and Danger: An Analysis of the Concepts of Pollution and Taboo* (London: Routledge).

Economic Research Service (2001), 'Briefing Room: Rice Background', Economic Research Service, USDA.

Ellstrand, N., Prentice, H. and Hancock, J. (1999), 'Gene Flow and Introgression from Domesticated Plants into their Wild Relatives', *Annual Review of Ecology and Systematics* 30, 539–63.

Fowler, C. and Mooney, P. (1990), *Shattering: Food, Politics, and the Loss of Genetic Diversity* (Tucson, AZ: The University of Arizona Press).

Fox, J. (2003), 'Puzzling Industry Reponse to Prodigene Fiasco', *Nature Biotechnology* 21, 3–4.

Freudenburg, W. and Gramling, R. (1994), *Oil in Troubled Waters: Perceptions, Politics, and the Battle Over Offshore Drilling* (Binghamton, NY: SUNY Press).

Geniella, M. (2004), 'Mendocino County Voters Ban Biotech Crops', *Santa Rosa Press Democrat*.

Gettman, J. and Armentano, P. (1998), 1998 Marijuana Crop Report: An Evaluation of Marijuana Production, Value, and Eradication Efforts in the United States.

Gottweis, H. (1998), *Governing Molecules: The Discursive Politics of Genetic Engineering in Europe and the United States* (Cambridge, MA: MIT Press).

Gould, F. (1998), 'Sustainability of Transgenic Insecticidal Cultivars: Integrating Pest Genetics and Ecology', *Annual Review of Entomology* 43, 701–26.

Hajer, M. (1995), *The Politics of Environmental Discourse: Ecological Modernization and the Policy Process* (Oxford: Oxford University Press).

Hamburg, L. (2003), 'Farmers, Families Fight for a GMO-free Mendocino', *Bullhorn*, Mendocino Organic Network.

Hay, I. (2000), *Qualitative Research Methods in Human Geography* (Oxford: Oxford University Press).

Heffernan, W. (1999), 'Biotechnology and Mature Capitalism', *11th Annual Meeting of the National Agricultural Biotechnology Council*, Lincoln, NE.

Henson, D. (2001), 'Asserting Democratic Control of Food Agriculture', *By What Authority? Program on Corporations, Law and Democracy*, Program on Corporations, Law and Democracy.

Howard, T. and Rifkin, J. (1977), *Who Should Play God? The Artificial Creation of Life and What it Means for the Future of the Human Race* (New York: Delacorte Press).

Kapuscinski, A. (2002), 'Controversies in Designing Useful Ecological Assessments of Genetically Engineered Organisms', in Letourneau, D. and Burrows, B. (eds), *Genetically Engineered Organisms: Assessing Environmental and Human Health Effects* (New York: CRC Press).

Kapuscinski, A. and Hallerman, E. (1990), 'AFS Position Statement: Transgenic Fishes', *Fisheries* 15, 2–5.

Kelso, D. (2004), 'Genetically Engineered Salmon, Ecological Risk, and Environmental Policy', *Bulletin of Marine Science* 74, 509–28.

Kenney, M. (1986), *Biotechnology: The University-Industrial Complex* (New Haven: Yale University Press).

Kloppenburg, J. (2004), *First the Seed: The Political Economy of Plant Biotechnology* (Madison: University of Wisconsin Press).

Krimsky, S. (1982), *Genetic Alchemy: The Social History of the Recombinant DNA Controversy* (Cambridge: MA, MIT Press).

Krimsky, S. and Plough, A. (1988), *Environmental Hazards: Communicating Risks as a Social Process* (Dover, MA: Auburn House Publishing Company).

Letourneau, D. and Burrows, B. (eds) (2002), *Genetically Engineered Organisms: Assessing the Environmental and Human Health Effects* (New York: CRC Press).

Losey, J. (1999), 'Transgenic Pollen Harms Monarch Larvae', *Nature* 399, 214–215.

Lucas, G. (2004), 'Biotech's Initiative's Foes Dig in Deep: $150,000 Donated to Bury Mendocino Crop-ban Measure', *San Francisco Chronicle*, San Francisco, CA.

Mccarthy, A. (2006), 'Why Sonoma County's Measure M Failed', *Terrain*, Ecology Center.

Mcgaughey, W., Gould, F. and Gelernter, W. (1998), 'Bt Resistance Management', *Nature Biotechnology* 16, 144–146.

Mendocino County Historical Society (1996), 'The Mills of Mendocino', *Mendocino County Historical Society*, Ukiah, CA.

Merrill, J. and Culp, C. (2001), 'Fetzer Wines Announce Opposition to GE Grapes', Greenpeace News Update.

Miller, B. (2000), *Geography and Social Movements: Comparing Antinuclear Activism in the Boston Area* (Minneapolis: University of Minnesota Press).

Mills, L. (2002), *Science and Social Context: The Regulation of Recombinant Bovine Growth Hormone in North America* (Montreal: McGill-Queen's University Press).

National Research Council (2002), *Environmental Effects of Transgenic Plants* (Washington, DC: National Academy of Sciences).

Nestle, M. (2003), *Safe Food: Bacteria, Biotechnology, and Bioterrorism* (Berkeley: University of California Press).

Peet, R. and Watts, M. (eds) (2004), *Liberation Ecologies: Environment, Development, and Social Movements* (London: Routledge).

Quist, D. and Chapela, I. (2001), 'Transgenic DNA Introgressed into Traditional Maize Landraces in Oaxaca, Mexico', *Nature* 414, 541–543.

Rifkin, J. (1998), *The Biotech Century* (New York: Putnam).

Rissler, J. and Mellon, M. (1996), *The Ecological Risks of Engineered Crops* (Cambridge, MA: MIT Press).

Segarra, A. and Rawson, J. (2001), 'StarLink Corn Controversy: Background', in Economic Research Service (ed.) (Washington, DC: Congressional Research Service).

Smith, N. (1992), 'Contours of a Spatialized Politics: Vehicles and the Production of Geographical Scale', *Social Text* 33, 54–81.

Snow, A., Andow, D., Gepts, P., Hallerman, E., Power, A., Tiedje, J. and Wolfenbarger, L. (2004), 'Genetically Engineered Organisms and the Environment: Current Status and Recommendations', *Ecological Applications* 13, 279–86.

Somers, T. (2004), 'Voters Shun Bioengineered Foods, but not Everyone in Quircky Northern California Region Favors Shutting Out Science', *San Diego Union-Tribune.*

Swyngedouw, E. (1997), 'Excluding the Other: The Production of Scale and Scaled Politics', in Lee, R. and Wills, J. (eds), *Geographies of Economies* (London: Arnold Press).

Union of Concerned Scientists (2003), 'Pharm and Industrial Crops: The Next Wave of Agricultural Biotechnology', Washington, DC.

Williams, R. (1989), *Resources of Hope* (London: Verso).

Wright, S. (1994), *Molecular Politics: Developing American and British Regulatory Policy for Genetic Engineering* (Chicago: University of Chicago Press).

PART 4
Contested Production of Environmental Science, Law, and Knowledge

Chapter 10

Regional Power and the Power of the Region: Resisting Dam Removal in the Pacific Northwest

Eve Vogel

Introduction

During the New Deal, the Columbia and Snake River systems were conceived as a single river basin unit, linked in identity and destiny with a Pacific Northwest region consisting of Washington, Oregon, Idaho, and western Montana. Together, the Columbia River basin and the three-and-a-half-state Pacific Northwest region were seen to constitute a particularly inclusive and democratic, socially and environmentally beneficial, regional scale for environmental planning and decision making. This regional-scale geographic framing has shaped Columbia and Snake River management thinking and decision making ever since.

This chapter examines how the historical fixing of the regional scale and these regional territories as the *right* scale and territories in which to frame questions about Columbia and Snake River management has directed and limited possibilities for river and salmon management for the last seven decades. I build on detailed archival research, extensive reading of published and unpublished secondary sources, my own personal experience with Columbia Basin fish and wildlife planning and policy analysis,[1] and numerous conversations with policy makers and interested parties. Conceptually, the chapter highlights the fact that fixed scales of environmental decision making inevitably have embedded content that can constrain and drive environmental policy choices, even decades after their fixing. It also reveals, though, that fixed scales are neither static nor autonomous. They evolve, and have leverage points for policy change even in the face of their continued power.

Scrutinizing one recent controversy, I show why the long-fixed regional scale of thinking and decision making in relation to the Columbia and Snake Rivers led to a convoluted decision in 2000 rejecting a plan to help endangered Snake River salmon by breaching four dams on the lower Snake River. The same analysis begins to suggest an alternate political strategy that might be more effective for those who advocate breaching the lower Snake dams.

---

1   I interned for a total of about half a year at the Northwest Power Planning Council in 2000 and 2001. My understanding of Columbia River politics grew in large part out of this experience, and from the many conversations I was able to have with policy-makers from federal agencies, states and tribes. This essay, however, does not in any way represent the views of Council (now the Northwest Power and Conservation Council).

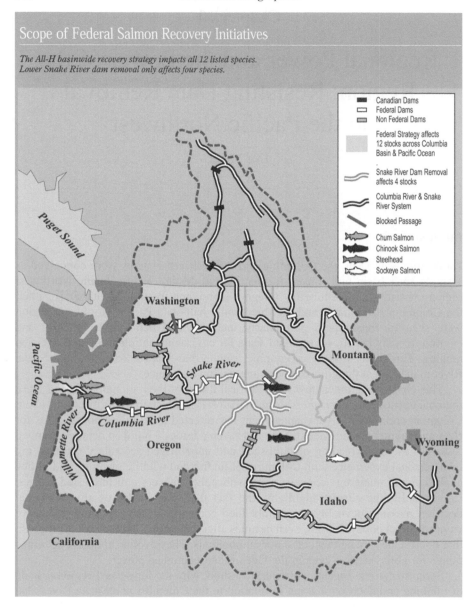

Figure 10.1 Regional-scale framing of a proposal to breach the lower Snake River dams to help Snake River salmon. Why look at the entire Columbia Basin to consider a question about the four lower Snake River dams? And what is the unlabelled dark grey area? The answers reveal the importance of a long-fixed scale and territory in Columbia River management and politics

Source: Federal Caucus (2000), 'Salmon and Our Future,' Citizen Update: Conservation of Columbia Basin Fish 4 (Summer): 2.

The chapter proceeds as follows. In the next section, I outline a way to unpack and historicize fixed scales of environmental decision making. Turning to the case study, I begin by uncovering the long-fixed scale behind the 2000 decision against breaching the lower Snake dams. I then go back in time to examine the origins of this fixed scale of decision making, to uncover its territorial, institutional, and ideological content, and its critical structural vulnerability. Next, I show how political pressure applied to this vulnerability has repeatedly effected limited but significant evolution in policy priorities, even in the face of the regional scale's continuing hegemony. Returning to the recent past, I show why the long-fixed regional scale of decision making led to a convoluted decision in which policy-makers rejected breaching the four Snake River dams. I suggest an alternative political strategy that might win removal of these dams. In the chapter conclusion, I return to the broader conceptual point, emphasizing that analyses of fixed scales of environmental decision-making can reveal both constraints on political options, and also leverage points for political change.

Unpacking and Historicizing Fixed Scales of Environmental Decision Making

As Brown and Purcell (2005) recently summarized succinctly, geographic scales are both fluid and fixed. The burgeoning literature on the politics of scale, though, has focused attention mainly on one side of this dichotomy: the fluidity of scales. It has by now been demonstrated that in myriad places and times scales have been contested, jumped, shifted, reconstructed, and interconnected (e.g. N. Smith 1993; Swyngedouw 1997; K. Cox 1998; Brenner 2001). Many have argued that recognizing these kinds of fluid processes of space, place and scale can open up opportunities for political challenge and change (for particularly strong arguments on this point see Allen 2004; Massey 2005).

But how do we reconcile this constructivist, dynamic, and relational notion of scales, and the desire for political challenge and change, with the recognition that scales can also be fixed? Seemingly most all the works that emphasize the fluidity of scales give a half-sentence nod to the fact that a scale or system of scales can be institutionalized in important ways. Such fixed scales shape and constrain the geographies in which people think, act, and relate to one another; the areas in which social, economic and political institutions are conceived, bounded and organized; and the ways individuals and a host of organizations represent and interact with the physical environment and non-human species (for welcome examinations of fixed scales see Zimmerer 2000; Haas this volume). Yet the fixed aspect of scales is much less explored. Without further exploration, we are left to think that when scales are fixed, politics are closed.

Yet simply revealing the existence of long-fixed scales is a politically important step. Much like long-established national territories and notions of place (Agnew and Corbridge 1995; Jess and Massey 1995), long-fixed scales are too often unexamined. Fixed scales of *environmental* regulation, management and decision making are perhaps especially unquestioned because they often draw on ideas of "natural" boundaries and systems (Haas this volume illustrates this point). Exposing them can bring them into the realm of the political, so that the way they frame environmental decision making and policy options is called publicly into question.

Uncovering the content of fixed scales of environmental decision making can open political discourse further. To fix a scale of environmental regulation and decision making for a particular issue or jurisdiction is likely to fix, at least roughly, a set of other things as well: territory, level of government, central policies and political relationships, fundamental values and goals. Each of these can shape and limit policy options. By unpacking and historicizing the dominant scales in which environmental problems are conceived and organized we can reveal this embedded content. Most important is to look at their moment of *construction*, as this is when territories, values, goals, and political relationships become embedded; and their moment of *institutionalization*, when scales become structured within particular organizations, territories, and legal and governmental forms (this is adapted from ideas of historical institutionalists e.g. Skocpol 1992; Thelen and Steinmo 1992; see Haas's chapter in this volume for an excellent illustration of this kind of examination).

Exposing the content of fixed scales of environmental decision making can also open up political possibility, even in the face of their continuing dominance, by exposing points of vulnerability which can be leveraged to effect policy change. Fixed scales should not be understood as static, immovable forces with determinate outcomes, but rather as path-dependent institutions that must be maintained actively by agents, and can evolve, change and be challenged, but only within certain constraints. Nor are fixed scales autonomous. Like all scales, they are connected with other scales and other axes of social power. These fluid aspects of scale inevitably create vulnerabilities that can sometimes threaten the continuance of a fixed scale of environmental management or decision making. Target these, and policies and politics supported even by powerfully fixed scales may change.

Over time, although a fixed scale of environmental decision making will remain fixed in its core shape and content, it can evolve in important ways. As it narrows its priorities or participants, or incorporates new ones, these changes inevitably influence environmental and social outcomes. Only by investigating the developmental history of fixed scales of decision making can one fully understand the ways they shape particular decisions at specific moments in time.

The Shady Regional Scale in the 2000 Decision not to Breach the Lower Snake River Dams

Figure 10.1 is a map of federal salmon recovery initiatives, published a few months before a December 2000 decision not to breach four federal dams on the lower Snake River in southeastern Washington to help endangered and threatened salmon. There are at least three features in the map that are unnoted in either the map's title or its legend. First, this map of 'federal salmon recovery initiatives' shows only those endangered and threatened salmon runs which spawn within the Columbia Basin— even though there were also, at the time this map was published, salmon listed under the federal Endangered Species Act in coastal Oregon, Washington, and California, as well as proposed for listing in Maine (National Marine Fisheries Service 1999a;

NOAA 1999). Second, the highlighted portion of the Columbia Basin stops not only at the international border but also at certain state borders. It is worth observing here that the highlighted area does not correspond to the salmon's range, as salmon cannot reach southern Idaho or any part of Montana. (The absolute limits to salmon migration are shown by the long grey bars which indicate dams impassable by salmon.) Third, there is an unlabelled dark shaded area that extends slightly to the east and the west of the highlighted portion of the Columbia Basin. Together, highlighted and shaded areas encompass the states of Washington, Oregon, Idaho and western Montana. There is a fourth emphasis that is explicit, but which bears noting as well: though the deliberation initially began focused on salmon in the Snake River basin, by the time this map was published, several salmon species in the portion of the Columbia Basin outside the Snake sub-basin had been listed under the Endangered Species Act as well, and the decision making had expanded to include them.

These four emphases reveal a geographical framing of the Snake River dams decision that neither the text of the map, nor the federal recovery documents with which it was associated, explain clearly: the unit of analysis was that portion of the Columbia Basin highlighted in the recovery map; and the core area for participation in the decision was the territory encompassing Washington, Oregon, Idaho, and western Montana. Although the decision was a federal government decision, the federal agencies – nine of them – organized themselves and their analyses mainly within these two regions or territories. They also sought input from these territories, tapping especially the governments of these four states and Columbia Basin Native American tribes (Federal Caucus 1999, 2000b).

While the two linked territories which framed the Snake River dams decision were often left unlabelled, the regional *scale* of the decision was emphasized repeatedly. Decision-makers frequently referred to the need for 'regional input,' 'regional participation', or for 'the region to decide' about the priorities for salmon and the lower Snake River dams (e.g. Brinckman 1998; Federal Caucus 2000b; Kempthorne et al. 2000). There was an implication that only a regional scale would successfully group the relevant social, economic and ecological needs, interests, and processes. Ironically, other geographic frames were suggested by interest groups that could also have been considered 'regional' in scale, but they were not called regional. Environmentalists emphasized the geography of the Snake River's Salmon and Clearwater River sub-basins (shown in central and northern Idaho in Figure 10.1). They argued that since these sub-basins have excellent habitat for salmon to spawn and rear, the critical factor for Snake River salmon must be the bottleneck in the lower Snake River (Save Our Wild Salmon Coalition 2000). The Chamber of Commerce in Lewiston, Idaho – the endpoint of a navigation corridor made possible by the lower Snake dams – focused on the entire journey of the Snake River salmon, from river tributaries through the lower Snake *and* lower Columbia Rivers, out into the ocean for hundreds of miles. Deflecting attention from the lower Snake, they asked in full-age newspaper ads, 'What about the other 4,900 miles?' (Lewiston Chamber of Commerce 1999). These were valid alternative framings for a question about Snake River salmon and Snake River dams. Nonetheless, their proponents did not affix to them the label 'regional' to help legitimize them. 'Regional' meant only the linked pair of regions shown on the federal recovery map.

I argue that this specific regional-scale geographical framing influenced the regulatory questions federal managers asked, and the policy options they entertained. Ultimately, it supported a decision not to breach the lower Snake River dams.

Mansfield and Haas (2006) show that different 'scale frames' can be used by politically interested parties to change the analytical focus in environmental conflicts. In the lower Snake River dams decision, though, the specific regional-scale decision making was not simply the product of a political manoeuvre for economic beneficiaries of the Snake River dams to trump environmental concerns. Federal agencies seemed genuinely concerned to meet broad environmental and social goals, and arguably, they did so. If the goal was to avoid impacts to dams and dam beneficiaries altogether, the larger-scale frame proposed by the Lewiston Chamber of Commerce would have been far more effective. In contrast, a Columbia Basin focus kept the pressure on Columbia and Snake River dams. While federal agencies declined to breach the lower Snake dams, they called for a huge array of projects throughout the basin, including expensive changes in water flows from and over multiple dams. They even kept dam breaching on the table as a backup in case other measures did not work (Federal Caucus 2000a; National Marine Fisheries Service 2000). If the Columbia Basin frame protected economic beneficiaries of the lower Snake River dams, it did so only with considerable expense, more pervasive if less dramatic regulatory restrictions, and limited long-term security. The final decision, then, was not so much one-sided as enormously convoluted. Many argued that breaching the lower Snake dams, together with some key investments to protect beneficiaries of these four dams, would have provided a simpler and more sustainable solution for environment and economy alike (e.g. Blumm et al. 1998).

These contradictions reflect the fact that regional-scale, three-and-a-half-state Pacific Northwest, Columbia Basin-focused frame had its own inertia, independent of the decision about the lower Snake River dams. On one level, the reason is simple, and follows straightforwardly from the Endangered Species Act. The decision not to breach the lower Snake River dams was made a part of the more encompassing 2000 Biological Opinion. The 2000 Biological Opinion, like several before it and since, grew from a request for 'consultation' under the Endangered Species Act by the three federal agencies which collectively operate the Federal Columbia River Power System (the agencies are the Bonneville Power Administration, Army Corps of Engineers and Bureau of Reclamation). The Federal Columbia River Power System operates 29 dams on the Columbia and Snake Rivers as one system (Bonneville Power Administration et al. 2001), and so it made sense to consider them a single 'agency action' under the Endangered Species Act. The three-and-a-half-state Pacific Northwest is the territory which receives preferential access to federal Columbia River power (*Pacific Northwest Consumer Power Preference Act* 1964), so it was logical that for a consultation concerning the Federal Columbia River Power System, this territory would get special input.

But to accept this as the whole answer is to assume that the 'single system' character of the Federal Columbia River Power System, and the three-and-a-half-state Pacific Northwest's preferential access to federal Columbia River power, need no explanation; and that other aspects of regional decision making follow automatically from these two geographical facts. But both parts need unpacking. There is no self-evident reason why the Federal Columbia River Power System is operated as one system in the first place – especially because many dams in the Columbia Basin are *not* a part of this system, while two dams on the separate Rogue River basin *are* a part of the system (Bonneville Power Administration et al. 2001); and because no other river basin in the United States except the Tennessee has a unified system of federal dams and hydropower. Nor is it apparent why Columbia River power is provided preferentially within a specific territory, and why that territory consists of Washington, Oregon, Idaho, and western Montana. Even if one accepts the regional power system and its preference region as givens, it is unclear why regional-scale decision making should confer special legitimacy in relation to broader questions about the Columbia River basin and its salmon, or about four dams in a very limited area within the 'region'. Finally, it is not apparent why regional decision making should aim to address and meet a wide and fairly inclusive range of social and environmental concerns, and yet lead to policy contortions to protect four dams.

The roots of this regional-scale geographical frame go deeper still. The geographic framing that was used in the Snake River dam decision is only one recent manifestation of the long hegemony of regional-scale thinking and decision making in the Columbia and Snake Rivers.

Unpacking and Historicizing the Columbia River-centered Pacific Northwest Region

Scalar construction: The embedded content in regional-scale Columbia River management

Regional-scale management of the Columbia River grew out of a historically unique constellation of three pairs of contrasting factors: technological development and the unique natural topography of the Columbia Basin; regional political mobilization versus top-down national directives; and regionalist ideals and shared self-interested ambitions for Columbia River development (see Vogel 2007 for a fuller elaboration of this history).

Prior to the New Deal, most people dealt with the Columbia River on their own local level, pursuing disparate and often competing ambitions for economic development. By the early 1930s, though, civic and business leaders in Washington, Oregon, Idaho and Montana knew that the Columbia River system had the greatest hydropower potential in the country, that newly matured long-distance transmission meant hydropower dams could provide power even to distant populations, and that the Army Corps of Engineers was preparing a basin-wide Columbia River development plan. When in early 1933, newly elected President Roosevelt and Congress approved Bonneville and Grand Coulee Dams on the Columbia River and Fort Peck Dam on the upper Missouri in Montana, and began to promote state and regional planning around the country, people from Seattle to Billings began to think about banding together to lobby for more federal works. In late 1933, representatives from Washington, Oregon, Idaho, and Montana came together with the regional representative of the new Public Works Administration to form the Pacific Northwest Regional Planning Commission (PNWRPC) (Bessey 1933, 1934).

The initial PNWRPC lacked any clear notion of the special legitimacy of the regional scale, a defining regional character, a unifying environmental resource, or shared vision other than economic development. Certainly, it saw Columbia River development as within its purview, but it initially spread its attention widely, attending to hundreds of public works proposals throughout its four-state area (Dana 1933; Bessey 1934; PNWRPC 1934). Ironically, a clearer regional conception would come to the Pacific Northwest primarily by way of national planners. In 1935, the PNWRPC's parent agency in Washington DC prepared a report on 'regionalism,' drawing from the ideals and analyses of several scattered groups of regionalist thinkers. These regionalists held that organizing society and economy at the 'regional' or subnational scale could reinforce a culturally rich, naturally caring society of small urban and lively rural communities and well-tended environmental resources (for an overviews see Dorman 1993; Meinig 2004). The draft report strongly supported organizing resource, economic and social planning and development at the regional scale, but, building from an impressively sophisticated regional geography analysis suggested dividing a 'Columbia Basin' region east of the Cascade Mountains from a west coast 'Pacific Northwest' region (Figure 10.2). This threatened to separate the Grand Coulee Dam and future Columbia River dam sites from the major west coast cities which hoped to benefit from the dams' power (Bessey 1935; NRC 1935).

In response, in a report prepared concurrently with the national regionalism report, the PNWRPC embraced the national planners' regionalist ideals and regional geography analysis methods but developed its own analysis to defend a different regional territory. Its only concession was to allow eastern Montana to join a Great Plains region. Using extensive data, the PNWRPC argued that Washington, Oregon, Idaho, and western Montana constituted what it called a 'homogeneous' region,[2] linked most centrally by the connecting influence of the Columbia River. The purpose of regional-scale coordination was to promote Columbia River development, and with that development, to advance a shared good life throughout this Pacific Northwest region that would be based on inclusive planning, wide sharing of resources, thoughtful balancing of rural and urban needs, and protection of a clean, bountiful natural environment (NRC 1936). Implicit, of course, was the assumption that the Pacific Northwest region would retain both right and responsibility for the Columbia River.

This, then, was the origin of regional-scale management of the Columbia River. I call this regional conception the 'Columbia River-centered Pacific Northwest'.

---

2   Hartshorne (1939) had yet to introduce the notion of functional regions.

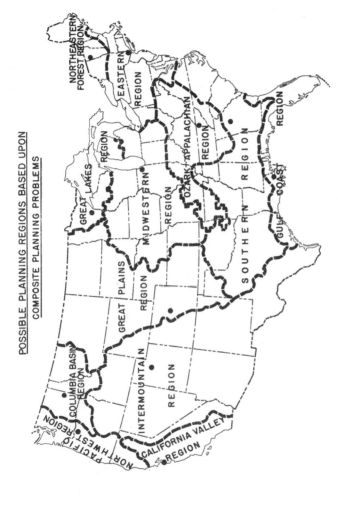

Figure 10.2 National planners' 1935 proposed regionalization of the country. This regionalization threatened to separate the fruits of Columbia River development from the cities of Seattle, Tacoma and Portland. It was this that prompted the articulation of a Pacific Northwest region that included all of Washington, Oregon, Idaho and western Montana, in which the Columbia River was a unifying tie

Source: National Resources Committee (1935), Regional Factors in National Planning and Development (Washington, DC: United States Government Printing Office): 166.

Institutionalized and narrowed regionalism through the Bonneville Power Administration

In the same report, the PNWRPC made recommendations for 'the policies and organization which should be provided for planning, construction, and operation' of Bonneville Dam and other public works in the Columbia Basin. Having asserted its territory's claim to Columbia River planning, the PNWRPC recommended a regional power transmission and marketing agency for Columbia River power that should be regionally based and whose purpose should be regionally and publicly defined. Power should be transmitted across the region and sold at low cost, so even marginalized people and communities like farm families and remote rural areas could benefit (NRC 1936) (see Figure 10.3).

From the time it was proposed, a regional Pacific Northwest power agency drew much more controversy than the PNWRPC, for both proponents and opponents recognized the real material, economic and political power it might wield. The bill which passed through Congress in 1937 was narrowed further to a power agency responsibly only for the power from Bonneville Dam (*Bonneville Project Act* 1937; Dick 1973; Funigiello 1973).

Out of these modest beginnings, the Bonneville Power Administration (BPA) grew and expanded over the next several years to become a fully regional agency encompassing much of the PNWRPC's three-and-a-half-state Pacific Northwest, and taking on many of that region's ideals. It built a regional transmission grid which closely followed the PNWRPC's plan, interconnecting Bonneville and Grand Coulee Dams and sending spur lines into northern Idaho. It offered rock-bottom electric power rates and helped to exand rural electrification in the three-and-a-half-state Pacific Northwest. It positioned itself as a kind of regional Chamber of Commerce, working with small towns and government agencies of all governmental levels to promote the Pacific Northwest as a destination for energy-intensive industries. And at the same time as all of this, it promoted protection of the Columbia Gorge and other scenic areas, and coordinated with fisheries agencies to compensate for impacts caused by the dams (Davis 1945a, 1945b; Ogden 1949; Bonneville Power Administration 1980).

Before long, BPA was *the* core regional institution in the Columbia River-centered Pacific Northwest. In 1943, the PNWRPC died with its national parent agency, at the hands of a Congress increasingly hostile to FDR's national planning (Clawson 1981). Now, the Columbia River-centered Pacific Northwest region narrowed. The focus of regional coordination became electric power, with other resources receiving only secondary attention. The vision of the good life and shared prosperity was largely reduced to a goal of wide distribution of inexpensive electricity and the promotion of widespread economic development. Intra-regional ties derived not from the interdependence of rural and urban, society and resources, but from a shared advantage in wider national and international markets, guarded with collective jealousy.[3] The region's economy and politics were tied indelibly to the Columbia River, but to the *developed* river, and to the dams which would soon multiply to become the Federal

---

3 Markusen (1987) suggests that shared economic advantage or disadvantage is usually the motivation for lasting regional political blocs in the United States.

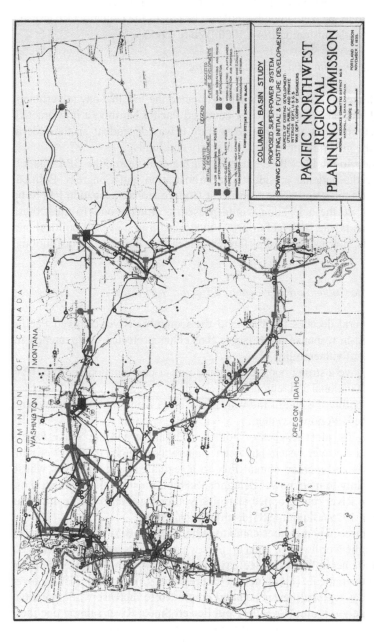

Figure 10.3 The PNWRPC proposed a regional power agency. The BPA followed this grid closely, adopting also the PNWRPC's regional conception and portions of its regionalist vision

Source: National Resources Committee (1936), Regional Planning, Part I–Pacific Northwest (Washington, DC: United States Government Printing Office): 40.

Columbia River Power System. Still, there continued to be a sense that the goal of regional coordination was ultimately the good life for both rural and urban areas throughout the three-and-a-half-state area, and that this as tied to protecting the river's other resources, to helping make the benefits of Columbia River resources widely available to all, and to the continuing bounty and beauty of the Columbia River itself (Bonneville Power Administration 1944, 1980; White 1995).

The structural vulnerability of a multi-state region

The fates of both the PNWRPC and the BPA reflected the fact that any government institution which put the Columbia River-centered Pacific Northwest into practice had a key vulnerability: as an inter-state agency it was dependent on the federal – that is, national – government for legal authority and funds.

As the sole heir to the PNWRPC's beneficent regional vision, the BPA was left alone to appease the national presidential administration and Congress. Political support within the four-state Pacific Northwest Congressional delegation had to be nearly unanimous, so the delegation would present a united front. To meet these political requirements, the BPA had to prove itself useful to powerful political and economic interests, not just poor rural farmers. Such was the price of institutional survival.

The evolution of the Columbia River-centered Pacific Northwest

Over the next several decades the BPA and the Columbia River-centered Pacific Northwest evolved in response to pressures that either protected or threatened its fundamental point of vulnerability.

World War II gave a strong boost to regional power coordination and the BPA. The coincidence of national and regional interests during the agency's youth enabled it to thrive. Even before the end of the war, many more dams were authorized and assigned to join the BPA power system.

By the mid-1970s, most reaches of both the Columbia and the Snake Rivers were dammed, and the river system had been inalterably changed from a river with high-velocity flows and marked annual flux to a regulated series of lakes whose flows changed mainly to meet power demands. Salmon populations declined; most notably, the Snake River runs dropped alarmingly after the four lower Snake dams were completed in 1975 (Blumm 1983; Lichatowich 1999).

In the 1970s, three new pressures came together to force BPA and the Columbia River-centred Pacific Northwest to open up to new participants, and to broaden their priorities to include heretofore excluded environmental concerns, Native American tribes and state governments. These new pressures were (a) a looming BPA power supply deficiency as regional power demand continued to increase while potential dam sites had largely run out; (b) widely publicized Snake River salmon declines; and (c) a series of Native American court victories upholding tribes' treaty-reserved, never-relinquished fishing rights (Lee et al. 1980; Blumm 1983).

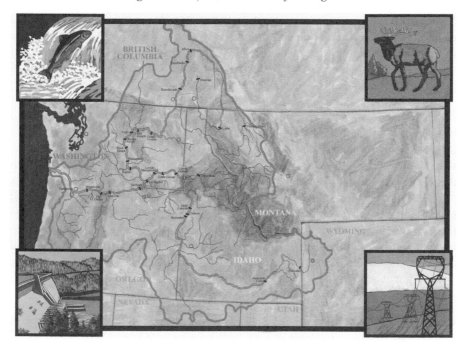

Figure 10.4   The Northwest Power Planning Council recast the old fixed scale of the Columbia River region as an inclusive river ecosystem

Source: Northwest Power Planning Council circa 2000. Image provided by the Northwest Power and Conservation Council, 2007.

In response to the power supply crisis, the four Pacific Northwest states called for an interstate compact that would give the state governments some control over BPA's power planning. The regional congressional delegation took their case to the national Congress. Once the proposal was in this national venue, fisheries interests found an ally in Michigan's Congressman Dingell, and were able to add in their concerns. The 1980 Northwest Power Act created a four-state council which undertook regional power and fish and wildlife planning, all of which the BPA was now expected to fund (*Pacific Northwest Electric Power Planning and Conservation Act* 1980; Blumm 1983; Bodi 1995). In effect, the act created a late twentieth century version of the PNWRPC, but one with Congressional authorization, a reliable and bountiful funding source (BPA's power proceeds), and a much stronger mandate to address fish and wildlife needs. The four-state Northwest Power Planning Council's logo reflected this new environmentally inclusive vision of the region (Figure 10.4).

Pushed further by Endangered Species listings of Columbia Basin salmon beginning in 1991, the BPA by the mid-1990s was spending over one hundred million dollars per year on salmon and other fish and wildlife–much of that funnelled into the fish and wildlife agencies of the four Pacific Northwest states and the Columbia Basin tribes (Committee on Protection and Management of Pacific

Northwest Anadromous Salmonids 1996; ISG 2000). Though the Columbia River-centred Pacific Northwest still was organized around electric power and the BPA as the core of regional well-being, its goal of environmental well-being had grown and shifted to embrace natural ecosystems and abundant wild fish populations; and its participants now included state and tribal fish and wildlife agencies (Lee 1995).

But these new state, tribal and environmental participants and interests were not the only ones who could leverage pressure on BPA. In the 1980s, the agency had been forced to absorb the debt from a set of never-completed nuclear power plants it had promoted and backed financially (D. Pope 2008). Now, while annual fish and wildlife costs were enormous, debt payments were even greater. Citing BPA's debt to the US Treasury as evidence of failure, some in Congress pushed to privatize the BPA or end its regional preference system. The BPA was forced to raise rates, but did all it could to protect the Columbia River-centred Pacific Northwest's continued access to the nation's cheapest power (Kriz 1997; C. Pope 2001). In this political context, the most important priorities for the BPA and its innumerable regional beneficiaries and allies were to ensure BPA's political survival and its cheap rates, and with these, the wide regional benefits that depended on them.

Resisting Dam Removal to Protect the Columbia River-Centered Pacific Northwest

At first, the proposal to remove the lower Snake dams to help Snake River salmon had been framed within the geography of the Snake basin, or at least that portion accessible to salmon. By the mid-to-late 1990s, within this frame of analysis, dam breaching began to seem like the preferred option, supported by ecology, law, economics, and even politics (see Blumm et al. 1998 for a summary of these arguments).

But as dam breaching began to seem genuinely possible, regional-scale political interests began to organize, and to re-scale the whole question. The problem for interests at the scale of the Columbia River-centred Pacific Northwest region were two. First, breaching the four dams would be expensive, at least before salmon returns could lower other fish and wildlife expenditures, and since the dams were Federal Columbia River Power System dams, it was likely the cost would be borne by BPA. This meant power rates across the region would go up (Anderson et al. 1997; US Army Corps of Engineers 2002). Second, decision-makers realized that to take out federal dams would take an act of Congress, and they feared that if Congress were asked to take out four Federal Columbia River Power System dams, Congress might also open up a full deliberation about the BPA. Either way, breaching the lower Snake River dams seemed to threaten the region's core institution and economic resource—and with them, at least in the minds of many policy-makers, the broad goals the region is still seen to fulfill (Hughes 1999; Swisher 1999; Barnett 2000).

A regional-scale decision had two essential advantages: regional analysis allowed enough factors to be considered that a decision against dam breaching could be justified; and regional coordination and participation could encompass all the interests which benefit from the preferential access to Federal Columbia River Power. Thus, the deliberation was expanded to include thirteen Columbia Basin salmon species, nine of which were outside the Snake basin, and had been listed under the Endangered Species Act just in time for consideration, in 1997–1999 (National Marine Fisheries Service 1997, 1998, 1999b, 1999c, 1999d). The deliberation reviewed the effects not just of the lower Snake dams, but the entire Federal Columbia River Power System – and, to open up enough management options to make up for continued losses through the dams, the entire river basin watershed as well. A nine-agency constellation of federal agencies coordinated by BPA brought in representatives from the four states and the basin's tribes for consultation, and conducted fifteen regional public hearings around the three-and-a-half-state Pacific Northwest (Northwest Fisheries Science Center 1999; Federal Caucus 2000b; National Marine Fisheries Service 2000).

Salmon advocates recognized their greatest hope in influencing the decision process lay in 'jumping scales' (K. Cox 1998) outside of this regional nexus. Even they had initially balked at the potential repercussions; they recognized that nationalizing the issue could threaten the BPA, and with it, hundreds of millions of dollars of fish and wildlife funding. Fed up with the recalcitrance of federal agencies in Columbia Basin salmon policies, though, they took their case to Congress and the Clinton administration (Barker 1999; Swisher 1999). Together, environmentalists and Native American tribes won supportive editorials in national newspapers (e.g. *New York Times* 2000), calls for dam breaching from many Congresspeople, and the ear of Vice President and presidential candidate Al Gore, who, many maintained, was eager to protect his flanks against a Green Party candidate (e.g. Montana 2001).

Environmentalists' and tribes' national mobilization did influence the decision – but they did not force breaching of the lower Snake dams, nor did they change the primary geographic scale of the decision. The fear of rising power rates or worse, of losing the BPA and the Federal Columbia River Power System entirely, outweighed the pressure salmon advocates could bring to bear. The BPA and the Corps, Pacific Northwest congressional representatives, industry and utility groups argued vehemently and successfully against breaching the lower Snake dams (Nolan 1997; Hughes 1999; Swisher 1999; Brinckman 2000; C. Smith 2000).

The over-all result was a convoluted decision. The National Marine Fisheries Service declined to call for breaching the lower Snake River dams. It argued that improvements in the tributaries and the estuary could make up for continuing losses through the lower Snake, and that breaching the lower Snake dams would not help the many listed species outside the Snake Basin (see italicized comment in Figure 10.1). Instead, it found a wider suite of measures was merited. It called for even greater sums of money than were already being spent on fish and wildlife, and significant and innovative measures, including supplementation of wild fish stocks, which Columbia Basin tribes had long advocated, and more comprehensive and widespread tributary habitat protection and restoration (Federal Caucus 2000a; National Marine Fisheries Service 2000).

Ongoing and future dam politics

Since 2000, environmental groups have continued to promote breaching the lower Snake River dams as the single most important action that could be taken to help Snake River salmon (e.g. Save Our Wild Salmon Coalition 2005a). Federal agencies, for their part, supported by the Bush administration, have tried to push dam removal further from consideration. A federal court judge overturned the 2000 Biological Opinion. The National Marine Fisheries Service (now going by the name NOAA-Fisheries) parried with a new Biological Opinion in 2004 which argued the Federal Columbia River Power System dams should be considered part of the 'environmental baseline' (NOAA Fisheries 2004). This, too, was overturned, and a new Biological Opinion is due in May 2008 (*Columbia Basin Bulletin* 2008). Besides this court battle, environmental groups have focused on further building their case that the benefits that would be lost from breaching the lower Snake River dams are replaceable (Taxpayers for Common Sense et al. 2006); and on building national congressional support for breaching dams (Save Our Wild Salmon Coalition 2005b).

It may be, though, that the scale-jumping national-scale political effort is not the best strategy for those who aim to have the Snake River dams removed. The long history and staying power of the Columbia River-centred Pacific Northwest, combined with BPA's financial burdens and its ongoing political vulnerability, suggest that any effort to remove Federal Columbia River Power System dams that begins with national-scale mobilization is bound to meet resistance from major political leaders in the Pacific Northwest. Congress is unlikely to act in the face of strong resistance from the regional delegation. It may be that advocates of dam removal need to come back to the regional scale, and with their participation reinvigorate a robust and inclusive Columbia River-centred Pacific Northwest. If they ally *with* the BPA, Pacific Northwest utilities and other key BPA customers, they might be able to create a united regional political coalition which would push Congress on two fronts: *first*, to protect BPA and Pacific Northwest inexpensive power; and then, once that is assured, to get Congress to authorize breaching the lower Snake dams.[4]

Conclusion

The history of the conception, institutionalization, and evolution of the Columbia River-centred Pacific Northwest region, combined with its role in the 2000 decision about the lower Snake dams, shows that long-fixed scales of environmental decision making can have profound influence on environmental conflicts, even decades after their inception.

In this case, the fixing of the regional scale for Columbia River management meant not simply establishing a scale, but also pairing a particular three-and-a-half-state political territory with the river's management; imbuing this regional scale territory, and its ties to the Columbia River, with both a set of social and environmental ideals and a shared self-interest in Columbia River development; and institutionalizing these ideals and this self-interest into a government agency with a specific topical focus and an inbuilt political vulnerability to national scale politics. Only by seeing that *all* of these things became embedded in regional scale

---

4   Discussion with environmental advocates reveal that this change in strategy is beginning to take place. In fall 2007, the Northwest Energy Coalition, which is close allies with the Save Our Wild Salmon coalition, advertised a job for a 'Utility Outreach Project Coordinator' for precisely this purpose (S. Weiss, personal communication, 2007).

management of the Columbia River can one begin to understand the paradoxical ways the 2000 Snake River decision played out. The decision about the dams turned against breaching the lower Snake dams, but not because interests grouped at this scale cared nothing for salmon or catered to a few dam beneficiaries, but because breaching the dams was seen as a potential threat to the entire regional power system, the most fundamental regional benefit of all. With dam breaching off the table, decision-makers had to call for a host of other measures which would benefit salmon and their human advocates. These collectively added up to great expense and limited security for dam beneficiaries.

This Pacific Northwest case study illustrates the power of and need for historical analysis of fixed scales of environmental regulation and decision making. In our enthusiasm to point out the many ways geographic scales can be fluid, we must not be blind to the existence and power of fixed scales of environmental regulation and decision making. But neither should we see fixed scales of environmental regulation or decision making as immobile, untouchable or autonomous. By historicizing them to uncover their embedded content, finding their leverage points, and tracing their evolution, we can both expose the way long-fixed scales favor particular priorities and constellations of interests, and begin to find political possibility even in the face of scalar inertia.

Acknowledgements

Archival research was conducted thanks to an EPA STAR fellowship and the able assistance of Monika Bilka. Huge thanks go to my advisor Alec Murphy for inspiring brainstorming and feedback, and to John Shurts of the Northwest Power and Conservation Council for helping me learn and think about Columbia River history and politics. For amazing personal support as I wrote and edited this chapter I thank Richard Turk and Ken Pendleton. The book editors are much appreciated for their generous time and constructive criticism on earlier drafts.

References

Agnew, J. and Corbridge, S. (1995), *Mastering Space* (London: Routledge).
Allen, J. (2004), 'The Wherabouts of Power: Politics, Government and Space,' *Geografiska Annaler, Series B: Human Geography* 1, 19–32.
Anderson, W. et al. (1997), *Fish and Wildlife Recovery in the Pacific Northwest: Breaking the Deadlock: A Draft Analysis by the Northwest Power Planning Council Staff* (Portland, OR: Northwest Power Planning Council), <http://www.nwcouncil.org/Library/1997/breaking/>.
Barker, E. (1999), 'Supporters of Breaching Go National,' *Lewiston Morning Tribune*, 21 October, 9A.
Barnett, J. (2000), 'Senate Candidate Commands Clout in Northwest,' *The Oregonian*, 26 October, A16.

Bessey, R.F. (1933), 'Minutes of Meeting after Umatilla Rapids Hearing at Pendleton,' 19 September National Archives and Records Administration, Pacific Alaska Region, RG 187, Box 6.

—— (1934), 'Minutes of First Meeting of Pacific Northwest Regional Planning Commission Held in Mr. Dana's Office,' 10 January. National Archives and Records Administration, Pacific Alaska Region, RG 187, Box 10.

—— (1935), 'Letter to Marshall Dana, Chairman, District No. 11, National Resources Committee – Confidential,' 10 July. National Archives and Records Administration, Pacific Alaska Region, RG 187, Box 22.

Bird, J. et al. (eds) (1993), *Mapping the Futures: Local Cultures, Global Change* (London: Routledge).

Blumm, M.C. (1983), 'The Northwest's Hydroelectric Heritage: Prologue to the Pacific Northwest Electric Power and Conservation Act,' *Washington Law Review* 58, 175–244.

Blumm, M.C. et al. (1998), 'Saving Snake River Water and Salmon Simultaneously: The Biological, Economic, and Legal Case for Breaching the Lower Snake River Dams, Lowering John Day Reservoir, and Restoring Natural River Flows,' *Environmental Law* 28: 4, 997–1054.

Bodi, L. (1995), 'The History and Legislative Background of the Northwest Power Act,' *Environmental Law* 25, 365.

Bonneville Power Administration (1944), *Pacific Northwest Opportunities: Preliminary Studies of Bonneville Power Administration with Indications of Basic and Related Programs of Other Agencies* (Portland, OR: Taylor and Co., Inc.).

—— (1980), *Columbia River Power for the People: A History of the Policies of the Bonneville Power Administration* (Portland, OR: Bonneville Power Administration).

Bonneville Power Administration et al. (2001), *Federal Columbia River Power System*, Brochure (Portland, OR: Bonneville Power Administration), <www.bpa. gov/power/pg/fcrps_brochure_17x11.pdf>.

Bonneville Project Act (1937), 16 USC, §832.

Brenner, N. (2001), 'World City Theory, Globalization, and the Comparative-Historical Method: Reflections on Janet Abu-Lughod's Interpretation of Contemporary Urban Restructuring,' *Urban Affairs Review* 37: 1, 104–147.

Brinckman, J. (1998), 'Consensus Critical to Salmon Efforts,' *The Oregonian*, 16 October, B6.

—— (2000), 'Unreleased Federal Plan Calls for Dam Breaching,' *The Oregonian*, 18 November, A1.

Brown, C.J. and Purcell, M. (2005), 'There's Nothing Inherent About Scale: Political Ecology, the Local Trap, and the Politics of Development in the Brazilian Amazon,' *Geoforum* 36, 607–624.

Clawson, M. (1981), *New Deal Planning: The National Resources Planning Board* (Baltimore: Resources for the Future, The Johns Hopkins University Press).

Columbia Basin Bulletin (2008), 'Redden Grants Biop Extension to May 5; 2008 Hydro Ops in the Works', *Columbia Basin Bulletin*, 1 February, <www.cbbulletin. com>.

Committee on Protection and Management of Pacific Northwest Anadromous Salmonids (1996), *Upstream: Salmon and Society in the Pacific Northwest* (Washington, DC: National Academy of Sciences).

Cox, K. (1998), 'Spaces of Dependence, Spaces of Engagement and the Politics of Scale, Or: Looking for Local Politics,' *Political Geography* 17: 1, 1–23.

Cox, K.R. (ed.) (1997), *Spaces of Globalization: Reasserting the Power of the Local* (New York: Guilford Press).

Dana, M.N. (1933), 'Letter to Charles W. Eliot, 2nd, Executive Officer, National Planning Board, Federal Emergency Administration of Public Works,' 29 August. National Archives and Records Administration, Pacific Alaska Region, RG 187, Box 1.

Davis, L. (1945a), 'History of the Bonneville Power Administration, November 1937 to September 1939' (Portland, OR: Bonneville History Project, unpublished report).

—— (1945b), 'History of the Bonneville Power Administration, September 1939 to January 1942' (Portland, OR: Bonneville History Project, unpublished report).

Dick, W.A. (1973), *Visions of Abundance: The Public Power Crusade in the Pacific Northwest in the Era of J.D. Ross and the New Deal*, Dissertation (Seattle, WA: University Of Washington).

Dorman, R.L. (1993), *Revolt of the Provinces: The Regionalist Movement in America, 1920-1945* (Chapel Hill: University of North Carolina Press).

Federal Caucus (1999), 'Salmon and Our Future,' *Citizen Update: Conservation of Columbia Basin Fish* 1 (Fall).

—— (2000a), *Conservation of Columbia Basin Fish, Final Basinwide Salmon Recovery Strategy* (Portland, OR: Federal Caucus).

—— (2000b), *Conservation of Columbia Basin Fish, Final Basinwide Salmon Recovery Strategy, Volume 3: Regional Coordination and Public Involvement* (Portland, OR: Federal Caucus).

Funigiello, P.J. (1973), *Toward a National Power Policy: The New Deal and the Electric Industry, 1933–1941* (Pittsburgh: University of Pittsburgh Press).

Gunderson, L.H. et al. (eds) (1995), *Barriers and Bridges to the Renewal of Ecosystems and Institutions* (New York: Columbia University Press).

Hartshorne, R. (1939), *The Nature of Geography: A Critical Survey of Current Thought in the Light of the Past* (Lancaster, PA: The Association of American Geographers).

Hughes, J. (1999), 'Three Senators Unite in Opposing Dam Removal,' *The Associated Press State and Local Wire*, 10 June.

ISG (Independent Scientific Group) (2000), *Return to the River: Restoration of Salmonid Fishes in the Columbia River Ecosystem* (Portland, OR: Northwest Power Planning Council) <http://www.nwcouncil.org/library/return/2000-12.htm>.

Jess, P. and Massey, D. (1995), 'The Contestation of Place,' in Massey and Jess (eds).

Kempthorne, D. et al. (2000), 'Recommendations of the Four Governors of Idaho, Montana, Oregon and Washington for the Protection and Restoration of Fish in the Columbia River Basin', <www.nwcouncil.org/library/2000/4governors.pdf>.

Kriz, M. (1997), 'Power Hungry,' *The National Journal* 29: 10, 448.

Lee, K.N. (1995), 'Deliberately Seeking Sustainability in the Columbia River Basin', in Gunderson et al. (eds).

Lee, K.N. et al. (1980), *Electric Power and the Future of the Pacific Northwest* (Seattle, WA: University of Washington Press).

Lewiston Chamber of Commerce (1999), 'The Other 4900 Miles' (poster advertisement).

Lichatowich, J.A. (1999), *Salmon without Rivers: A History of the Pacific Salmon Crisis* (Washington, DC: Island Press).

Mansfield, B. and Haas, J. (2006), 'Scale Framing of Scientific Uncertainty in Controversy over the Endangered Steller Sea Lion,' *Environmental Politics* 15: 1, 78–94.

Markusen, A. (1987), *Regions: The Economics and Politics of Territory* (Totowa, NJ: Rowman and Littlefield).

Massey, D. (2005), *For Space* (London: Sage Publications).

Massey, D. and Jess, P. (eds) (1995), *A Place in the World? Places, Cultures and Globalization* (Oxford, UK: Oxford University Press).

Meinig, D.W. (2004), *The Shaping of America: A Geographical Perspective on 500 Years of History, Volume 4: Global America, 1915–2000* (New Haven: Yale University Press).

Montana, C. (2001), 'Dam Breaching a Political Hot Potato', *Indian Country Today* 23 August.

National Marine Fisheries Service (1997), 'Endangered and Threatened Species: Listing of Several Evolutionarily Significant Units (ESUs) of West Coast Steelhead', Final rule, 62 Federal Register 43937.

—— (1998), 'Endangered and Threatened Species: Threatened Status for Two ESUs of Steelhead in Washington, Oregon, and California', Final rule; Notice of determination, 63 Federal Register 13347.

—— (1999a), 'Designated Critical Habitat: Critical Habitat for 19 Evolutionarily Significant Units of Salmon and Steelhead in Washington, Oregon, Idaho, and California', Final rule, 65 Federal Register 7764.

—— (1999b), 'Endangered and Threatened Species: Threatened Status for Three Chinook Salmon Evolutionarily Significant Units (ESUs) in Washington and Oregon, and Endangered Status for One Chinook Salmon ESU in Washington', Final rule, 64 Federal Register 14308.

—— (1999c), 'Endangered and Threatened Species: Threatened Status for Two ESUs of Chum Salmon in Washington and Oregon', Final rule; notice of determination, 64 Federal Register 14508.

—— (1999d), 'Endangered and Threatened Species: Threatened Status for Two ESUs of Steelhead in Washington and Oregon', Final rule; Notice of determination, 64 Federal Register 14517.

—— (2000), *Biological Opinion: Reinitiation of Consultation on Operation of the Federal Columbia River Power System, Including the Juvenile Fish Transportation Program, and 19 Bureau of Reclamation Projects in the Columbia Basin* (Seattle, WA: National Marine Fisheries Service).

New York Times (2000), 'Saving the Snake River Salmon', *New York Times*, 2 April, 14.

NOAA (National Oceanic and Atmospheric Administration) (1999), 'Endangered and Threatened Species; Proposed Endangered Status for a Distinct Population Segment of Anadromous Atlantic Salmon (*Salmo Salar*) in the Gulf of Maine', Proposed rule, Notice of public hearing, 64 Federal Register 62627.

NOAA Fisheries (2004), *Endangered Species Act – Section 7 Consultation: Biological Opinion* (Seattle, WA: NOAA's National Marine Fisheries Service (NOAA Fisheries)).

Nolan, R.S. (1997) 'Columbia River Alliance for Fish, Commerce and Communities', Digital Studios (updated 22 March) <http://www.cyberlearn.com/cra.htm>.

Northwest Fisheries Science Center, National Marine Fisheries Service (1999), *An Assessment of Lower Snake River Hydrosystem Alternatives on Survival and Recovery of Snake River Salmonids, Appendix to the US Army Corps of Engineers' Lower Snake River Juvenile Salmonid Migration Feasibility Study* (Seattle, WA: Northwest Fisheries Science Center, National Marine Fisheries Service).

NRC (National Resources Committee) (1935), *Regional Factors in National Planning and Development* (Washington, DC: United States Government Printing Office).

—— (1936), *Regional Planning, Part I–Pacific Northwest* (Washington, DC: United States Government Printing Office).

Ogden, D.M., Jr. (1949), *The Development of Federal Power Policy in the Pacific Northwest*, Dissertation (Chicago: University of Chicago).

Pacific Northwest Consumer Power Regional Preference Act (1964), 16 USC, §837.

Pacific Northwest Electric Power Planning and Conservation Act (1980), 16 USC, §839.

PNWRPC (Pacific Northwest Regional Planning Commission) (1934), 'Proceedings of the First Pacific Northwest Regional Planning Conference', 5–7 March, Portland, Oregon (Portland, OR: Pacific Northwest Regional Planning Commission).

Pope, C. (2001), 'BPA Advantages Criticized in Report: Market Rates and End of Subsidies Sought for Federal Power Supplier That Has Been Good to NW Economy', *Seattle Post-Intelligencer*, 31 May.

Pope, D. (2008), *Nuclear Implosions: The Rise and Fall of the Washington Public Power Supply System* (Cambridge, UK: Cambridge University Press).

Save Our Wild Salmon Coalition (2000), *Press Release: Draft Columbia/Snake River Salmon Plan Won't Save Salmon from Extinction*, <http://removedams.org/library/lib-detail.cfm>.

—— (2005), 'About Us', Save Our Wild Salmon Coalition, <http://removedams.org/about/>.

—— (2005), 'Mission and History', Save Our Wild Salmon Coalition, <www.wildsalmon.org/about/mission-history.cfm>.

Skocpol, T. (1992), *Protecting Soldiers and Mothers: The Political Origins of Social Policy in the United States* (Cambridge, MA: Belknap Press of Harvard University Press).

Smith, C. (2000), 'Testimony', Subcommittee on Fisheries, Wildlife, and Drinking Water Committee on Environment and Public Works, US Senate, 20 November <http://epw.senate.gov/107th/smi_1100.htm>.

Smith, N. (1993), 'Homeless/Global: Scaling Places', in Bird et al. (eds).

Steinmo, S. et al. (eds) (1992), *Structuring Politics: Historical Institutionalism in Comparative Analysis* (Cambridge, UK: Cambridge University Press).

Swisher, L. (1999), 'Friends of Fish Teaming up with BPA Foes', *Lewiston Morning Tribune*, 16 May, 4C.

Swyngedouw, E. (1997), 'Neither Global nor Local: "Glocalization" and the Politics of Scale', in Cox (ed.).

Taxpayers for Common Sense et al. (2006), *Revenue Stream: An Economic Analysis of the Costs and Benefits of Removing the Four Dams on the Lower Snake River*, Save Our Wild Salmon Coalition, <www.wildsalmon.org/library_files/revenuestream8.pdf>.

Thelen, K.A. and Steinmo, S. (1992), 'Historical Institutionalism in Comparative Politics', in Steinmo et al. (eds).

US Army Corps of Engineers (2002), *Lower Snake River Juvenile Salmon Migration Feasibility Report and Environmental Impact Statement* (Walla Walla, WA: US Army Corps of Engineers) <http://www.nww.usace.army.mil/lsr/final_fseis/study_kit/studypage.htm>.

Vogel, E. (2007), 'The Columbia River's Region: Politics, Place and Environment in The Pacific Northwest, 1933-Present', Dissertation (Eugene, OR: University of Oregon).

White, R. (1995), *The Organic Machine: The Remaking of the Columbia River* (New York: Hill and Wang).

Zimmerer, K.S. (2000), 'Rescaling Irrigation in Latin America: The Cultural Images and Political Ecology of Water Resources', *Ecumene* 7: 2, 150–75.

Chapter 11

Law of Regions: Mining Legislation and the Construction of East and West

Johanna Haas

Every day, we see the effects of law on the landscape and the environment. Much writing within geography focuses on measuring and understanding the impacts of particular environmental laws, with much less attention paid to a critical understanding of legal geographies (Mustafa 2001). Some geographers have studied effects of international environmental agreements (see e.g. Marchak 1995; Soroos 1999; Demeritt 2001). Many studies linking legal causes to geographic effects have been done in the areas of civil rights and the urban form (see e.g. Mitchell 1997; Blomley 1998; Mitchell 1998). Environmental justice literature examines the social and environmental effects of legal processes. These studies look at the ties of risk and harm to class and race (Beck 1992; Bowen, Salling et al. 1995), engage scale as a social construct (Kurtz 2002; Kurtz 2003), and place space in a functional relationship with both ideas and the law (Pulido 2000).

Still, most of this work analyses the *effects* of the law, rather than looking into its *causes*. In this article, I hope to expand on these analyses, examining the law from both ends and showing linkages between the *causes* and the *effects* of law. Law is more than simply a mechanism through which geographic and environmental results are created, although it does act as that, having immense effects on landscapes and livelihoods. Rather, the law intertwines with the construction of regions as well as their institutionalization. Legal process and legal history add an institutional dimension to the social construction of scale as expressed within political ecology. Through analysis of the constructed meaning and discourse created through legal writing, it is possible to trace the interactions of the imaginations of place with the materialization of place, which in turn shapes a new imagination or strengthens an existing one.

This chapter uses critical analysis of legal discourse to examine the language creating and enforcing legislation through an examination of the Surface Mining Control and Reclamation Act of 1977 (SMCRA) and the debates over its passage throughout the 1970s. Because legislative history looks beyond the word of the law to examine the documents that explain the passage of the law (Breyer 1992), my method is to use legislative history to find ideas and beliefs that shaped the institutional imagination and thus Congressional action which created material results through the mining landscape. Congressional documents from SMCRA debates preserve not only debate over the bill, but also record external influences that shaped thought processes behind the bill forming the institutional imagination of Congress.

This paper will trace the multi-directionality of social construction as embodied in the legal process and show how examination of the legal process adds a dimension to the social construction of region and scale. First, I will lay out the concept of the institutional imagination as one way to examine social construction within legal systems. Second, I will examine the strengths and weaknesses of study at the intersections of law and geography. Finally, I will illustrate the processes involved in these constructions through an examination of the passage and substance of surface mining law.

Institutional Imagination

Extractive industries, by their nature, affect regions in a spatially uneven fashion. With coal mining, this divide breaks down into a devastated East and a wide-open West. The Appalachian East represented the worst of eastern devastation as absentee landowners had extracted wealth in coal and timber, and moved on leaving the land bare and useless (Caudill 1962; Eller 1982; Rasmussen 1994; Lewis 1998). The East is seen as a 'used' landscape (McSweeney and McChesney 2004). The West, meanwhile, is wrapped in ideas of the wide open frontier (Worster 1992; Cronon 1997). Western resources, abundant and unclaimed, sat ready to be harnessed by humans and used for progress (Worster 1985). Nature-based industries, and increasing harvests, were the key to wealth and development (Prudham 1998). While the Appalachian people of the East were portrayed as victims of the environmental harms caused by careless industry, the people of the mountain West were portrayed as stewards of a bountiful environment. These ideas are recorded in the historical legal process surrounding the development of surface mining legislation.

Mining reclamation captured the institutional imagination of Congress during debates over the SMCRA, which shaped (and was shaped by) the geography of mining in the United States. The regional differences formed a regional imaginary, a set of ideas that I call *institutional imagination*, that Congress shared surrounding regions, industry, and environmental ideals. The legal use of geographical ideas can cause and shape geographical change (Delaney 1995) as it was shaped by the changes which happened before. The final surface mining bill resulted in many more restraints on eastern mining than on western mining, institutionalizing growth in western coal paralleling shrinking production in the East. SMCRA itself defines the regions, drawing a line down the middle of the United States at the 100th Meridian, setting different requirements for mining reclamation east and west of that line.

The institutional imagination of Congress created a new, institutionalized reality with wide-ranging effects. This institutional imagination shapes federal law, which incorporates spatial and geographic ideas to create spatial and geographic results and through that process, a legal topic becomes naturalized and legitimated (Jessop 2002). Law does more than simply influence existing regions; it shapes them and then institutionalizes that shape, giving it a permanence that few other processes can achieve. Through the process of legalization, an idea gains entry into the realm of governance, and through that, gains a large set of enforcement mechanisms: public and private, formal and informal (Goodwin and Painter 1996). Law invokes powerful institutionalizing forces: attention of multiple regulatory agencies, citizen monitoring through court suits, increased scientific study, and more. The combined result of these forces are an entry into a cultural lexicon that divides actions between 'right' and 'wrong' where legal action is right and illegal action is wrong. Passage of a law becomes part of society's expectations, and through that, becomes legitimated at a variety of levels, including enforcement of both the goals of the law itself and the ideas that it embodies.

Legal Geographies

Law, although highly concerned with place and region when defining territory and jurisdiction, resists being linked to the social sciences. The traditional view of the law is as a science (Sutherland 1967), built from universally applicable laws which can be tested through falsifiability. In this viewpoint, law is a system that provides a mechanism and institution into which the facts can be applied to reach one unique and correct disposition (Hoeflich 1986). The law, as a unique science, is therefore insulated from other disciplines in the same way that one would not apply principles from psychology to geology. The law, therefore, looks inward for rules, theories, and models, rooted in 'natural law' (Burton 1988). Through this, law claims internal rationality which is key to avoid criticisms of arbitrariness and nihilism that would cause it to be seen as invalid, and challenge its very existence (Fiss 1982).

Legal work utilizes a realist conceptualization of nature and society embodied in the law, tracing current law back to an earlier 'natural law,' or law as unchanging and eternal. Because of this underlying idea, legal research does little to examine possibilities within geography and social theory by constructing itself as insulated from social forces (Delaney 2001a, 2001b). Social construction is particularly problematic within legal thought, and while critical legal scholars have worked on realizing this alternative conceptualization of the law, their work is shunted to the background in policy-making and other mainstream legal applications. Opening the possibility that law might be a social construct risks destroying the idea of law as a closed system that will logically and institutionally give correct results in every case, as if correctness is predetermined (Blomley 1994). Once a legal decision is made, it is considered to not only be law from that point on, but to have been law *as it always was* (Friedman 1985; Clark 1992). Simply put, if law is internally changeable, it ceases to be 'law'. Law may be a mirror of culture, but more importantly, it is a culture in and of itself (Shamir 1996) and internal preservation of that culture requires a rejection of social construction.

Blomley (1994) examines how legal scholars claim that federal law is not only geographically neutral but also geographically neutralizing. Law is assumed to exist outside of society, providing rules that create uniformity and link all places together. In reality, neither law nor space is neutral, with both involving social forces and power relations. Law provides a link between power and space. Most legal analysis has a tendency to either overlook space (and place), or assume it is merely a backdrop (Ford 1994). Instead, ideas about space and place emerge through the legal process, expressed in court cases through both legal arguments and judgments, invoking issues of agency and power (Delaney 1995). Once in place, legal ideas and legal spaces enter power structures of law and government, becoming self-perpetuating. Space becomes the result of law, rather than a context for events or a territory to be contested (Ford 1994). Legal space is itself a set of practices, both social and legal, which confirm and enhance its existence as a space (Ford 1999).

SMCRA began with the idea of creating practices to regulate the effects of surface mining for coal, but more than that, it began with ideas of East and West being inherently different. Regions were implicitly understood to be the framework in which the law operated, and this institutional imagination shaped the law as it formed. Region has long been central to the discipline of geography. Since the 1980s, scholars have viewed the region as a dynamic process instead of a unit of analysis. This 'new regionalism' is a method of analysis that incorporates economic, political, cultural, and a variety of other contexts to illustrate how they work to create and define regions (Allen, Massey et al. 1998; MacLeod 1999; Paasi 2002). Recognition of region as construct allows for greater understanding of processes at a variety of scales, showing landscape as linked to a variety of political and economic practices (Walker 2003).

Legislative process can be one important factor in the creation of a new region as it can highlight and differentiate regional formation. The law provides a long-standing and formal process of social construction (Delaney 2001). In this case, legislation is a social construction of one particular regional formation, and gradually ideas of what constitutes a region become solidified as social facts, either widely accepted by the populace or contested. A variety of mechanisms flesh out law and make it into fact – the courts, the media, and business all enact processes that reinforce legal mechanisms. The law provides a particularly important method of social construction through its ability to institutionalize ideas, giving them permanence. As a region's definition emerges in the political process, it gains institutional thickness. The regions that emerge from Congressional processes take hard form. SMCRA regulated mining in east and west quite differently – as Congress saw different needs for each region. Through this, legislation created different institutionalized forms for mining processes in the different regions.

Institutional Imagination in Legislative Process

Congress spent many years debating surface mining legislation. This long process leaves a trail of documents recording thoughts as Congress debated, shaped, and changed a variety of bills. These documents record a distinction between East and West that existed in the institutional imagination, where the East was viewed as harmed by previous extractive activity and needing protection by the federal government while the West was seen as open for natural resource development and guarded by careful stewards. The process started with a pre-existing difference between East and West, which was understood by Congress in particular ways. However, in creating a law, Congress accomplished more than serving as a reflection of already existing conditions. By making these distinctions part of the law, Congress solidified them, giving them both greater permanence and the institutional support provided by the law.

Surface mining law, once institutionalized, had many wide-ranging effects. With SMCRA, Congress tackled the idea of regulation for methods during and reclamation after coal mining. This legalization spurred the creation of the Office of Surface Mining. It led other administrative agencies, both state and federal, to devote significant resources to monitoring and studying mining and its aftermath. Once agencies began acting, the expectations of citizens about mining were heightened, leading to environmental and community group activism over mining issues. However, as law operated differently in the West and East, these mechanisms and expectations developed differently, leading to different mining cultures.

History of the legislation

While the first surface mining bill was introduced in 1940, the serious push toward national surface mining control began in 1971, with several surface mining bills introduced each year following, until 1977 when one bill became law. These bills ranged from a total ban on surface mining to mild bills that closely mirrored existing state regulations. During this time, surface mining surpassed underground mining in total production – accounting for 60 per cent of the nation's coal production (Green 1997).

The 93rd Congress, in 1973–1974 saw the first major strip mine bill effort by Rep. Morris Udall, who represented a large mining region in southern Arizona, and became the major force behind the bill that eventually would become law. As the Chairman of the House Interior Committee (and brother of former Secretary of the Interior Stewart Udall), Udall was the land policy leader in the House (Miller 1991). The Interior Committee drafted a complex bill that looked much like other environmental regulation that Congress had recently enacted (H.R. 11500, 14 May, 1974). This bill spurred six days of floor debate in Congress, spawning over 100 amendments with argument breaking down along eastern and western lines. A rift developed between the two mining regions with different physical features, modes of mining, and methods of reclamation. Mining, which was physically easier in the West, was growing rapidly there, causing a strong reaction from eastern mining companies who wished to protect their market shares. In December 1974, a conference report passed both houses and was sent to the president for his signature. However, this final bill came out at the end of the legislative session and quietly sat on President Ford's desk until it met a pocket veto on 30 December, 1974, as he ignored it to death (Miller 1993).

The surface mining bill that came out of the 95th Congress was the result of many years of trial and failure. This time, Congress knew that the game had changed and a new president held the White House. Congress also knew that experts forecast great growth for coal in the West, where the coal was thick and economically attractive, particularly in the Northern Great Plains (1977 USCCAN at 611). These coals lay closer to the surface of the land and were easier to extract, leading to cheaper production (1977 USCCAN at 613). Western coal also had much lower reclamation costs than coal in the East (1977 USCCAN at 619). Through many provisions of the bill, SMCRA gave western mining a green flag. The mines could continue and grow, as long as they followed a few guidelines. President Carter signed the bill on 3 August, 1977.

Institutional Imagination

Institutional imagination of the West

The first bill, from 1940, followed the introduction of the first state laws to regulate surface mining. While some states had strong laws on paper, many states took a lax stance on enforcement. Western interests recognized federal regulation of strip mining was inevitable. Thus, they supported a bill to regulate mining instead of banning it, and favoured the weak model of existing state laws. Western mining companies objected to the federal government overseeing a regulatory plan, however, pragmatic mine operators realized that if federal regulation would be happening, it would be best to guide it instead of fighting it (Dyer 1980; Dragoo 1993).

Driven by these political strategies, Western interests in Congress worked to create an institutional imagination of the West that utilized the idea of land stewardship. Rep. Melcher from Montana stated, 'But Montana, and, I believe, the rest of the West, cherish our land to the extent that we do not want it strip mined without ironclad guarantees that the land will be restored resulting in successful and permanent vegetation' (118 Cong. Rec. 35,045, 1 October, 1972). In this manner, Congress began to differentiate effects of mining in the East and the West, and the imagination of East and West themselves. The debate over surface mining happened at the same time as the Sagebrush Rebellion, which along with its successor the Wise Use movement, were key movers in asserting the ideas of Western land stewardship (McCarthy 2002). Western interests attempted to assert their stewardship construction over that of the urban East which 'viewed the West as a wilderness playground that must be preserved and not developed' (Marzulla 1995: 238). Instead, they constructed the idea of the West as shaped, changed, and improved by careful human action in the name of progress (Worster 1985). Importantly, this progress was characterized as one based on a strong land ethic, aimed at preserving the West as a useful landscape (Marzulla 1995). Westerners themselves utilize these ideas, and discuss many different environmental situations referring to their ethic of stewardship and responsibility (Waage 2001).

Institutional imagination of the East

The popular press fed into the construction of a damaged East. Fears over existing environmental harms caused by steep-slope mining in Appalachia motivated the drafters of this bill to craft strict environmental protections for the East. Harry Caudill's books influenced many people at that time and the Congressional Record reflects that members of Congress passed around a copy of *Night Comes to the Cumberlands* (McElfish and Beier 1990). The book, by Kentucky lawyer Caudill, explores the devastation of strip mining, and ties it to the poverty of the region's residents (Caudill 1962). Some members of Congress even took a field trip to view devastation caused by surface mining in West Virginia and Pennsylvania (but took no such trip to view mines in the West). They saw direct damage to the landscape caused by highwalls, slag-piles, and subsidence. This view of Appalachian devastation combined with another set of images from President Johnson's War on Poverty, which presented Appalachian people as malnourished, unemployed, and living in shacks in remote 'hollers' (Weller 1965; Precourt 1983). The popular perceptions of Appalachia and Appalachian mining created an almost paternal need in Congress to protect the eastern mountains.

Combined institutional imaginations

The effect of this bill, causing migration of mining interests, was also understood within the institutional imagination. Congress was aware of the pre-existing migration of the mining economy. At every step, the decisions they made could boost or devastate an entire region of the country, and this outcome fit perfectly with the needs created within the institutional imagination. When working on SMCRA, Congress had a clear picture of the devastation caused by strip mining in the East. Meanwhile, Congress saw the West as a place where careful stewards would work to avoid similar harms. Thus, many of the substantive provisions of the act placed a greater burden on eastern mining companies than their western competitors, and the final bill resulted in more procedural restraints on eastern mining than on mining in the West. The resulting law affected the coal industry by speeding 'the development of western coal production because of its low cost and its low sulfur content, to the detriment of eastern coal producers ...' (Green 1997). This result was shaped by an institutional imagination that saw the need to protect the East while simultaneously seeing the West as capable of protecting itself.

Institutional Imagination in Legislative Content

This East/West imaginary shaped not only the legislative process, but also the content of the final law. Further, it resulted in a greater institutionalization of regional differences. Many of the substantive provisions of SMCRA operate specifically east or west of the 100th Meridian, while others regulated environmental phenomena found predominately in one area or the other. These provisions institutionalized different mining processes for each region, and, through that, created institutionalized understandings of the regions. The law turns regions into a fact for the people who live and work there, by enabling the variety of processes that reinforce regional differences. Each of the provisions was a result of a deliberative process, where Congress debated various outcomes and decided to enable the development of western coal mining while restricting mining development in the East.

While Congress created a flexible SMCRA for the West, they created a much stricter law for the East. Western mining companies were given choices in the methods they used and ways to avoid prohibitions on mining in certain areas. Meanwhile, the provisions applying to eastern areas set out detailed requirements for mining method in many situations and set unavoidable prohibitions on mining in problematic areas. Western results-based standards allowed for greater industry profit (and growth) than Eastern performance-based standards.

There are five major areas where SMCRA created stricter requirements for the East than for the West. Together, these provisions barred eastern mining companies from making decisions about the manner and method of mining. These strict provisions recognize harms that mining had previously done to the East and work to prevent them in the future. First, many of SMCRA's rules deal with difficulties of reclamation in steep and hilly areas where any disturbance of a slope can cause erosion problems. Most of SMCRA's steep slope provisions are inflexible, calling for specific methods of performance and lack location- and method-based flexibility. These rules lock eastern miners into specific methods of reclamation that add greatly to the costs of mining.

Second, Congress mandated specific methods for excess spoil disposal. Disposal of excess spoil (the dirt and rock removed to reach the coal) is a problem much more prevalent in Appalachia and other hilly areas than in flatlands. Third, SMCRA provides that miners must return mined lands to their approximate original contour. Congress also noted the economic impossibility of meeting the approximate original contour (AOC) requirements fully in Appalachia (1977 USCCAN at 654). Even realizing that AOC would be unachievable, Congress still enacted a rigid and detailed set of performance standards in the law itself – instead of flexible results standards.

Fourth, SMCRA requires the elimination of all highwalls. While this measure makes great environmental sense, the costs of doing so are very high. Reclamation on a highwall parallels to that on the walls of the big pit mines of the West. However, Congress mandated very different treatment of the two structures, allowing big pits while outlawing highwalls. Fifth, Congress created specific rules for construction of mine waste and water impoundments. SMCRA limited these impoundments as much as possible, mandating size, stability in construction, water quality and stability in water levels, through creating specific rules for the methods used, instead of leaving the decisions to best available practice and technology.

When comparing SMCRA in East and West, a pattern appears throughout the provisions of the law. Western mining companies were given the opportunity to plan different mining practices, and evaluated on the perceived results of those practices. Meanwhile, eastern mining companies faced strict performance-based standards, which mandated the methods used. There are four major ways that SMCRA institutionalized the idea of Western mining as stewardship. First, pre-existing state statutes in the West were similar to the federal law that Congress eventually enacted. Western mining companies had to change their mining processes less than their Eastern counterparts in order to comply with the federal law.

Second, SMCRA weakened the ban on mining in river valleys. In the West, the same valleys that hold vast coal reserves are also the focus of ranching and farming economies, and have considerable hydrologic importance to the arid West (Miller 1993). In order to mine in these areas, the operator needs to prove that the operation will not harm agriculture and ranching in the area and will not "materially damage" the water quantity or quality. Because of pressure from western states, Congress implemented this provision instead of a much stricter one that would have banned all mining on alluvial valley floors (1977 USCCAN at 714). Third, SMCRA eased

Western mining through another flexible ban on mining in sensitive areas. Congress created exceptions to provisions that set aside sensitive lands as unsuitable for mining, allowing for mining where companies held valid existing rights or mined in national forests without significant forest cover. Fourth, Congress treated the West differently and more stringently than the East in dealing with water issues. Mining in the West faces the problem of reclamation in a dry climate, as water issues become much more important when water is scarce and disturbances in the land structure have impacts on the hydrologic equilibrium. Each mine site, both in the East and West, must have a cumulative hydrologic impact assessment prepared, but in reality this only affects mining in the west. This provision requires western miners to plan carefully for not only the reclamation of the land but also reclamation of the water.

Thus, SMCRA created different rules for eastern and western mining operations. Western companies were cast in the role of stewards, and given flexible results-based standards to meet as well as choices of how to achieve the desired results. Eastern companies, meanwhile, were seen as untrustworthy through their responsibility for earlier mining devastation. Congress required them to follow strict laws and mandated performance-based standards that removed most of the choices they could make. In general, the western law has flexible results standards while the eastern law has strict performance standards. This final shape of SMCRA's provisions is a result of the institutional imagination about East and West.

Conclusions

SMCRA, along with advances in mining technology, has lead to enormous growth in western mining while eastern mining has lagged behind. By 2002, Wyoming produced 34 per cent of the total coal mined in the US, its highest year ever. West Virginia came second at 14 per cent, however its highest production year was 1947. Kentucky held third place with 11 per cent (National Mining Association 2003). SMCRA is certainly not the sole cause of these changes, as topography, technology, and other forces have favored western mining as well. However, SMCRA's provisions and the institutional imagination that they embody are one important part of this process. The mining economy had already started to migrate from the East to the West, and Congress' work in SMCRA codified and institutionalized that movement. I am not saying that environmental regulation of mining is undesirable, rather that the law created different material results in East and West.

Today, the landscape created by SMCRA is changing again. Mining is again expanding in the east, in a form called mountaintop removal mining, which is tied to a loophole the mining companies found in the Act allowing mountaintops to be removed without having to return them to the approximate original contour. This pairs with an agency reinterpretation of one word, while waste is not allowed to be placed in streams, fill is, and what was formerly labelled as waste is now labelled as fill, and the reinterpretation allows entire valleys to be filled in. Again, eastern mining is reaching the size and scope of western mining – but in a way that leaves behind an unusable and unrecognizable landscape.

Meanwhile, the dangers of mining are yet again entering the popular imagination, and yet again being seen as an eastern problem. On 3 January 2006, an underground coal mine in Sago, West Virginia exploded, killing twelve miners and causing a national uproar, especially as the details of their deaths and the safety measures not taken began to emerge. This tragedy was soon followed by several more mining deaths in West Virginia and Kentucky, making 2006 the deadliest mining year in recent history. These disasters, and the public outcry, have brought Congress to examine the issue of mining regulation again. Only time will tell what shape these regulations will take, but it is certain that institutional imaginations of place will emerge, shaping the legal process, and in turn being shaped by the laws that result.

Acknowledgements

This chapter is dedicated to the families of the twelve men who lost their lives at Sago, West Virginia on 3 January 2006, bringing the rest of us the coal that enables our daily lives.

References

(1977), Surface Mining Control and Reclamation Act (30 USC sec. 1201 et seq.), United States Code, Title 30.

Allen, J., Massey, D. et al. (1998), *Rethinking the Region* (New York: Routledge).

Beck, U. (1992), *Risk Society: Towards a New Modernity* (London: Sage Publications).

Blomley, N. (1994), *Law, Space, and the Geographies of Power* (New York: Guilford Press).

Blomley, N. (1998), 'Landscapes of Property', *Law and Society Review* 32: 3, 567–612.

Bowen, W.M., Salling, M.J. et al. (1995), 'Toward Environmental Justice: Spatial Equity in Ohio and Cleveland', *Annals of the Association of American Geographers* 85: 4, 641–663.

Breyer, S. (1992), 'On the Uses of Legislative History in Interpreting Statutes', *Southern California Law Review* 65, 845.

Burton, S. (1988), 'Judge Posner's Jurisprudence of Skepticism', *Michigan Law Review* 87, 710.

Caudill, H.M. (1962), *Night Comes to the Cumberlands: A Biography of a Depressed Area* (Boston: Little, Brown and Co).

Clark, G.L. (1992), 'The Legitimacy of Judicial Decisionmaking in the Context of *Richmond v. Croson*', *Urban Geography* 13.

Cronon, W. (1997), 'The Trouble with Wilderness: Or, Getting Back to the Wrong Nature', in C. Miller and H. Rothman, *Out of the Woods: Essays in Environmental History* (Pittsburgh: University of Pittsburgh Press), 28–50.

Delaney, D. (1995), 'The Boundaries of Responsibility: Interpretations of Geography in School Desegregation Cases', *Urban Geography* 15: 5.

Delaney, D. (2001a), 'Environmental Regulation: Introduction', in N. Blomley, D. Delaney and R.T. Ford, *The Legal Geographies Reader* (Malden, MA: Blackwell), 218–220.

Delaney, D. (2001b), 'Making Nature/Marking Humans: Law as a Site of (Cultural) Production', *Annals of the Association of American Geographers* 91: 3, 487–503.

Demeritt, D. (2001), 'The Construction of Global Warming and the Politics of Science', *Annals of the Association of American Geographers* 91: 2, 307–337.

Dragoo, D.A. (1993), 'West of the 100th Meridian: Implementation of the Surface Mining Control and Reclamation Act in the Western States', in U. Desai, *Moving the Earth: Cooperative Federalism and Implementation of the Surface Mining Act* (Westport, Conn: Greenwood Press), 167–182.

Dyer, B. (1980), *The Surface Mining Act in the West: A Challenge for Cooperative Federalism* (Lexington, KY: Council of State Governments).

Eller, R.D. (1982), *Miners, Millhands, and Mountaineers: Industrialization of the Appalachian South, 1880–1930* (Knoxville: University of Tennessee Press).

Fiss, O. (1982), 'Objectivity and Interpretation', *Stanford Law Review* 34, 739.

Ford, R.T. (1994), 'The Boundaries of Race: Political Geography in Legal Analysis', *Harvard Law Review* 107.

Ford, R.T. (1999), 'Law's Territory (A History of Jurisdiction)', *Michigan Law Review* 97.

Friedman, L.M. (1985), *A History of American Law* (New York: Touchstone).

Goodwin, M. and Painter, J. (1996), 'Local Governance, the Crises of Fordism and the Changing Geographies of Regulation', *Transactions of the Institute of British Geographers, NS* 21, 635–648.

Green, E.M. (1997), 'Colloquium on SMCRA: A Twenty Year Review: State and Federal Roles Under the Surface Mining Control and Reclamation Act of 1977', *Southern Illinois University Law Journal* 21, 531–551.

Hoeflich, M.H. (1986), 'Law and Geometry: Legal Science from Leibnitz to Langdell', *American Journal of Legal History* 30, 95.

Jessop, B. (2002), 'Liberalism, Neoliberalism, and Urban Governance: A State-Theoretical Perspective', *Antipode* 34: 3, 542–472.

Kurtz, H.E. (2002), 'The Politics of Environmental Justice as the Politics of Scale: St. James Parish, Louisiana, and the Shintech Siting Controversy', in Herod, A. and Wright, M.W. *Geographies of Power: Placing Scale* (Malden, MA: Blackwell), 249–273.

Kurtz, H.E. (2003), 'Scale Frames and Counter-scale Frames: Constructing the Problem of Environmental Injustice', *Political Geography* 22, 887–916.

Lewis, R.L. (1998), *Transforming the Appalachian Countryside: Railroads, Deforestation, and Social Change in West Virginia, 1880–1920* (Chapel Hill, NC: University of North Carolina Press).

MacLeod, G. (1999), 'Place, Politics, and "Scale Dependence": Exploring the Structuration of Euro-regionalism', *European Union and Regional Studies* 6, 231–253.

Marchak, P. (1995), *The Tropical Forestry Action Plan and Plantation Forestry. Logging the Globe* (Montreal and Kingston: McGill-Queens University Press), 194–216.

Marzulla, N.G. (1995), 'Property Rights Movement: How it Began and Where it is Headed', in Yandle, B. *Land Rights: The 1990s Property Rights Rebellion* (MD, Rowman and Littlefield).

McCarthy, J. (2002), 'First World Political Ecology: Lessons from the Wise Use Movement', *Environment and Planning A* 34, 1281–1302.

McElfish, J.M. and Beier, A.E. (1990), *Environmental Regulation of Coal Mining: SMCRA's Second Decade* (Washington, DC: Environmental Law Insitute).

McSweeney, K. and McChesney, R. (2004), 'Outbacks: The Popular Construction of an Emergent Landscape', *Landscape Research* 29: 1, 31–56.

Miller, C.W.J. (1991), *Stake Your Claim!: The Tale of America's Enduring Mining Laws* (Tucson: Westernlore Press).

Miller, R.O. (1993), 'The Surface Mining Control and Reclamation Act: Policy Structure, Policy Choices, and the Legacy of Legislation', in Desai, U. *Moving the Earth: Cooperative Federalism and Implementation of the Surface Mining Act* (Westport, Conn: Greenwood Press), 17–30.

Mitchell, D. (1997), 'The Annihilation of Space by Law: The Roots and Implications of Anti-homeless Laws in the United States', *Antipode* 29: 3, 303–335.

Mitchell, D. (1998), 'The Scales of Justice: Localist Ideology, Large-scale Production, and Agricultural Labor's Geography of Resistance in 1930s California', in Herod, A. *Organizing the Landscape: Geographical Perspectives on Labor Unionism* (Minneapolis: University of Minnesota Press), 159–194.

Mustafa, D. (2001), 'Colonial Law, Contemporary Water Issues in Pakistan', *Political Geography* 20, 817–837.

National Mining Association (2003), US Coal Production by State by Rank, National Mining Association, 2003.

Paasi, A. (2002), 'Bounded Spaces in the Mobile World: Deconstructing "Regional Identity"', *Tijdschrift voor Economische en Sociale Geografie* 93: 2, 137–148.

Precourt, W. (1983), 'The Image of Appalachian Poverty', in Batteau, A. *Appalachia and America: Autonomy and Regional Dependence* (University Press of Kentucky), 86–110.

Prudham, W.S. (1998), 'Timber and Town: Post-war Federal Forest Policy, Industrial Organization, and Rural Change in Oregon's Illinois Valley', *Antipode* 30: 2, 177–196.

Pulido, L. (2000), 'Rethinking Environmental Racism: White Privilege and Urban Development in Southern California', *Annals of the Association of American Geographers* 90: 1, 12–40.

Rasmussen, B. (1994), *Absentee Landowning and Exploitation in West Virginia, 1760–1920* (Lexington: University Press of Kentucky).

Shamir, R. (1996), 'Suspended in Space: Bedouins Under the Law of Israel', *Law and Society Review* 30: 2.

Soroos, M.S. (1999), 'Global Institutions and the Environment: An Evolutionary Perspective', in Vig, N. and Axelrod, R. *The Global Environment: Institutions, Law, and Policy* (Washington, DC: Congressional Quarterly Press), 27–51.

Sutherland, A.E. (1967), *The Law at Harvard: A History of Ideas and Men, 1817–1967* (Cambridge: Harvard University Press).

Waage, S.A. (2001), '(Re)claiming Space and Place Through Collaborative Planning in Rural Oregon', *Political Geography* 20, 839–857.

Walker, P.A. (2003), 'Reconsidering "Regional" Political Ecologies: Toward a Political Ecology of the Rural American West', *Progress in Human Geography* 27: 1, 7–24.

Weller, J.E. (1965), *Yesterday's People: Life in Contemporary Appalachia* (Lexington, KY: University of Kentucky Press).

Worster, D. (1985), *Rivers of Empire: Water, Aridity, and the Growth of the American West* (New York: Oxford University Press).

Worster, D. (1992), *Under Western Skies: Nature and History in the American West* (New York: Oxford University Press).

PART 5
Fraught Spatial Technologies and Knowledge Construction

Chapter 12

Mapping Boundaries, Shifting Power: The Socio-Ethical Dimensions of Participatory Mapping[1]

Jefferson Fox, Krisnawati Suryanata,
Peter Hershock, and Albertus Hadi Pramono

Introduction

The recent growth in the availability of modern spatial information technology (SIT) – geographic information systems (GIS), low-cost global positioning systems (GPS), remote sensing image analysis software – as well as the growth of participatory mapping techniques has enabled communities to make maps of their lands and resource uses, and to bolster the legitimacy of their customary claims to resources by appropriating the state's techniques and manner of representation (Peluso 1995). Since the publication of Hugh Brody's seminal work on mapping the lands of native Americans in the Canadian sub-Artic (1981), participatory mapping has enabled the successful demarcation of land claims that led to: the signing of treaties (e.g. Nisga'a); compensations for land loss (Native American, Maori); and formation of indigenous territory and government (e.g. Nunavut).

But, the impacts of widespread adoption of SIT at the local level are not limited to the intended objectives. Among the unintended consequences of mapping have been increased conflict between and within communities (Sirait et al. 1994; Poole 1995; Sterritt et al. 1999); loss of indigenous conceptions of space and increased privatization of land (Fox 2002); and increased regulation and co-optation by the state (Urit 2001; Majid Cooke 2003). Consequently, mapping technology is viewed as simultaneously empowering and disadvantaging indigenous communities (Harris

1 An earlier and shorter version of this chapter was published as an introduction to *Mapping Communities: Ethics, Values, Practice* (East-West Center, 2005) that documents the collection of case studies covered by our project. Yet another shorter communication was published as 'Mapping power: ironic effects of spatial information technology', in *Participatory Learning and Action* 54: 98–105 (2006). This chapter is based upon work supported by the National Science Foundation under Grant No. SDEST-0221912, the Rockefeller Brothers Fund, and the Ford Foundation (through the Jakarta Office). We would like to thank all the participants to the workshops for sharing their ideas. Any opinions, findings and conclusions or recommendations expressed in this material are those of the authors and do not necessarily reflect the views of the National Science Foundation (NSF), the East-West Center, or the University of Hawaii.

and Weiner 1998). Researchers working under the umbrella of Research Initiative 19 of the National Center for Geographic Information and Analysis (NCGIA) suggest that GIS technology privileges 'particular conceptions and forms of knowledge, knowing, and language' and that the historical development of the technology leads to 'differential levels of access to information' (Mark et al. no date). Rundstrom (1995) further suggests that GIS is incompatible with indigenous knowledge systems and separates the community that has knowledge from information (the 'product' of GIS application). Tensions thus exist between new patterns of empowerment yielded through SIT and broader social, political, economic, and ethical ramifications of the technology.

We submit that the tools, families of technologies, and practices associated with SIT use are value-laden and that deploying SIT will necessarily have ethical consequences. That is, the deployment of SIT will affect the constellations of values that distinctively shape any given society, its spatial practices, and its approach to reconciling conflicts or disharmony among competing goods or interests. We further submit that because the tools and technological families gathered under the rubric of SIT were not originally developed and produced in rural communities or among indigenous peoples in Asia, it will be in such settings that the tensions associated with SIT and its ironic effects are likely to be most apparent and potentially profound. To date, most research on the social and ethical implications of spatial information technology has been conducted in North America (Sieber 2000). Given the rapidity with which the use of SIT is becoming 'necessary', there is an urgent need to examine the implications of this technology – especially in rural settings and in less developed countries, as well as among indigenous groups.

Unintended Consequences of Mapping

This chapter and the research project on which it is based emerged out of common and yet distinct concerns among the authors that spatial information technologies – at least in certain contexts and at certain scales – can lead to consequences that raise important ethical questions. We identified three inter-related dimensions in which these consequences have manifested: in conflicts correlated with changing patterns of spatial perceptions and values; in competition related to knowledge and claims of resources; and in relation to structural or organization stresses at the institutional level. Our observation began with discussions in relation to the experiences of one author (Fox) with participatory mapping activities in Southeast Asia, where he has been working since 1983.

Based on a series of interviews conducted with villagers in the Ratanakiri Province in northeastern Cambodia between 1995 and 1997, Fox observed that mapping village boundaries resulted in changes in local practices that governed access and territoriality (Fox 2002). Prior to this mapping, villagers had a clear sense of their respective rights to ancestral lands, but specific boundaries between hamlets were not required unless cultivation areas from two hamlets met one another. In such situations, village elders of both communities might meet to decide the boundaries. Villagers respected others' rights because they believed that crossing another hamlet's swidden fields frequently would make the spirits unhappy and cause misfortune or death. In response to growing land pressures, however, a local leader convened village headmen together to map village boundaries. Subsequently, they began prohibiting people from other hamlets to use lands within the boundaries of the village – even in areas where they did not have to cross each other's fields (Fox 2002).

In another project, Fox observed that mapping raises critical questions on how map production, distribution, and ownership potentially consolidate control over spatial information. In 1997, a group of non-government organizations, community representatives and university researchers in Indonesia began to map land-use practices and boundaries of fourteen villages that lie in and around Wanggameti National Park in Sumba, Eastern Indonesia (Hardiono et al. 2005). Toward its completion in 2002, they brought the nearly completed maps back to villagers to assess their accuracy. At this time, villagers were also asked if they would grant permission to the project to distribute the maps. While villagers did not mind distributing the maps to the NGOs and university researchers, many objected to sharing the information that the maps contained with government agencies. Yet everyone realized that if maps were distributed to the NGOs, no one could guarantee that they would not fall into the hands of government officials. In order to retain some control of the information they provided, villagers decided to keep copies of the maps that they themselves could provide to organizations that sought them. Unfortunately, the villages' remote locations have limited practical utilizations of the maps for developing management plans for the park, leaving the multi-year mapping project largely futile (Hardiono et al. 2005). This controversy points to the important challenge of resolving the question of map ownership prior to initiating a participatory mapping activity.

Finally, working with two different NGO groups in Indonesia, Fox had experiences that raised questions about the impacts of participatory mapping activities on non-government organizations and their employees (Hardiono et al. 2005). In the early 1990s, a small NGO in Jakarta requested Fox to train one of their employees in GIS and related spatial information technologies, so that the NGO could integrate a spatial component into their work. A foundation grant was obtained to fund the training in the East-West Center's lab in Honolulu for several months, and to purchase equipment for a GIS lab in his NGO. Within several months of establishing the new lab, the director of the NGO fired the newly trained GIS specialist. Initially, Fox viewed the problem as a simple clash of personalities. The problem, however, was repeated several years later with another NGO, in which a newly trained GIS specialist and the leader of the NGO had a falling out, forcing the GIS specialist to leave the group.

We argue that adoption of SIT could alter the organizational hierarchy of an NGO, aligning staff members that are skilled with mapping technologies separately from the rest. Personnel skilled in spatial information analysis are still relatively scarce in many developing countries such as Indonesia. As participatory mapping approaches become popular, demand for such personnel increases. Organizations advocating environmental and community interests compete not only with each other but also with private mapping consultants, driving up not only the prestige, but also the salary structure of mapping and GIS technicians. Such a situation leads to tensions within NGOs, as well as a relatively fast turnover of spatial information specialists (Hardiono et al. 2005).

This chapter evinces our efforts to critically broaden reflection on such experiences and their implications for technology transfer and evaluation. Our analysis of these phenomena is informed by studies in technology and society that examine the interplay between technological development and the social institutions that shape its further deployment. Furthermore, we examine these issues from a political ecology perspective that situates the proliferation of SITs in the context of economic and political liberalization in many counties in Asia, exemplified by the rise of decentralization policies and community-based approaches since the 1980s (Brosius et al. 1998; Ribot 2002). These reforms have brought an explosion of new property claims and protectionist strategies in forests and other environments changing the very terms by which resources and environments are defined.

Tools, Technologies, and Ironic Effects

Critically assessing the impacts of SIT requires us to clarify the relationships between tools and technologies. Tools are products of technological processes. They are used by individual persons, communities, corporations, and nation-states– and they are evaluated based on their task-specific utility. If tools do not work, they are exchanged, improved, cannibalized, or discarded. In contrast, technologies consist of widespread patterns of material and conceptual practices that embody and deploy particular strategic values and meanings (Hershock 1999). Technologies are complex systems promoting and institutionalizing relational patterns aimed at realizing particular ends. Technologies cannot be value neutral, and do not occur in isolation from one another but in families or lineages (Shrader-Frechette and Westra 1997; Hershock 1999).

A hand-held GPS unit, for example, is a tool associated with SIT. Individuals using GPS units assess them in terms of their reliability, ergonomic design, technical specifications, and features. By contrast, SIT as a whole consists of a complex system of material and conceptual practices. They include: the extraction of raw materials; their manufacture into tools like GPS units, notebook computers, and satellites; the storage of information in massive, internet mediated databases; advertising and marketing these tools, the services associated with them, and the 'worlds' to which they provide access; the crafting of industry-specific regulatory and legal institutions; new patterns of expert testimony in legal contests over land-use; and, a reframing of the politics of development. As technology, SIT transforms the discourse about land and resources, the meaning of geographic knowledge, the work practices of mapping and legal professionals, and, ultimately, the very meaning of space itself.

There are two major implications of the tool/technology distinction. First, while we can refuse to use a tool, there are no clear 'exit rights' from the effects of heavily deployed technologies, even if individuals elect not to use the tools produced as part of that deployment. The concept of exit rights in discussions of technology and ethics invokes rights not to be subject to the use or effects of particular technologies and their associated tools. Serious questions arise regarding the possibility of exit rights with respect to technologies that are deployed at sufficient scale to make viable alternatives practically nonexistent. For example, although one can elect to not own or use a personal computer, computing technology is so widely deployed that it is not possible to avoid its effects. In practical terms, we have no exit rights from the computerized world.

Second, critical evaluation of a technology must go beyond assessing how well the tools specific to the technology perform, to examining the changes it brings about within and among societal systems and values. If viable exit rights do not exist for a technology, technologies can only be fully and effectively evaluated in terms of how they transform the quality of relationships constituting our situation as a whole. These relationships include those we have with our environment; with one another; with our own bodies; and with our personal, cultural, and social identities. In short, technologies must be evaluated in explicitly social and ethical terms.

Critical histories of technology deployment (see, for example Illich 1973; 1981) suggest that there are thresholds of utility for any given technology, beyond which conditions arise that make its broader and more intensive deployment practically necessary. That is, when a technology is deployed at sufficient intensity and scale, it effectively undermines the possibility of exercising exit rights with respect to the technology, generating problems of the type that only that technology or closely related ones can address. These distinctive patterns of ironic (or 'revenge') effects (Tenner 1996; Hershock 1999) have wide-ranging, systemic ramifications well outside the technology sector. For example, automotive transportation technologies were originally adopted to make transportation faster and easier, and to reduce urban pollution (from horse-drawn carriages). Their widespread adoption, however, transformed both environmental and social realities in ways that eventually generated problems – for example, inhospitable urban sprawl, traffic gridlock, and massive air pollution – that can only be addressed through more and better transportation (and transportation relevant) technology.

Ironic effects demonstrate the fallacy in assuming that what is good for each of us will be good for all. The individual user of tools is not, therefore, a suitable unit of analysis in critically assessing technologies. In addition, ironic effects argue for recognizing that the causality of technological impacts is fundamentally non-linear. Although new technologies are practically built from 'the ground up' by bringing together knowledge and materials in novel ways, once they are fully realized, the technology begins exerting 'downward causation' (Lemke 2000) on its component systems, bringing them into functional conformity with its own systemic needs.

Following this argument, once spatial information technologies cross the threshold of their utility, their use will become practically imperative and they will begin generating ironic or revenge effects that effectively command deploying these technologies at ever greater scales and intensities. While individual users may be benefited in anticipated fashions, the impacts at community level are less certain. More specifically, we submit that the widespread adoption of SIT will disadvantage small, local communities that – relative to other actors and stakeholders – have limited access to SIT as well as limited (material, conceptual, and professional) resources for making use of SIT in advocacy, legislative, and regulatory settings. Increased dependence on SIT will transform the relationships between human actors and their spatial environments in ways that correlate with loss of the indigenous spatial practices that were originally to be conserved through their deployment.

Workshop on SIT and Society

In order to test and further refine our ideas about the socio-ethical implications of SIT deployment, we convened a workshop in Chiang Mai, Thailand in June 2003. In planning and hosting the workshop we sought groups that have used SIT extensively in their community-based work. Altogether twenty-three participants that included officials from non-governmental organizations (NGOs), project staff members, and university researchers attended the weeklong discussion. They represented eight groups in seven countries (Cambodia, China, Indonesia, Malaysia, the Philippines, Thailand, and the United States). Workshop participants were introduced to key concepts for evaluating SIT in terms of its socio-ethical effects, including the concepts of exit rights and ironic effects. Participants then worked in small groups to reflect on their own experiences in grassroots implementation and deployment of SIT. These results were shared in plenary sessions and further developed and refined through group discussions. Discussions were guided by three interlinked and overlapping sets of questions, summarized in Table 12.1.

We first sought to understand the *social and political dynamics* that result in communities choosing to engage in mapping, focusing on the ways maps and mapping shift patterns of resource control. The second set of questions addressed the *socio-ethical implications* of mapping technologies and activities. Spatial information technologies have embedded within them values such as 'universality', 'objectivity', 'standardization', 'precision', and 'control' that have emerged in systemic relationship within the context of a particular historical/cultural experience. The introduction of these technologies into societies where these values have been neither prominent nor systematically integrated may have unexpectedly disruptive effects. The last set of questions examined the impacts of SIT on the *organizational dynamics of the non-governmental organizations (NGOs)* that introduce SIT into rural communities. We began with a position that the adoption of spatial information technologies by NGOs is problematic because of their social context, the potential for co-optation, and a lack of resources.

Following the workshop, participants were invited to prepare research proposals. After consultation with the authors/project leaders, seven groups were funded by a grant from the Rockefeller Brothers Fund. These groups spent the next year conducting research at their respective organizations and field sites. The groups reconvened with the authors in Honolulu in October 2004 to write papers based on what they learned from their research. These papers were published in a volume edited by Fox, Suryanata and Hershock (2005), which this chapter summarizes.

Table 12.1   Questions to guide discussions in the Chiang Mai workshop

Why maps? Enrollment and empowerment	• Why do communities engage in mapping? Local and extra-local reasons for communities to adopt SIT • Who got empowered? Against whom? • What are the processes by which empowerment occurs? • Who controls the maps? How do various actors decide how maps can be utilized?
Socio-ethical implications	• Are there any changes in conceptions of space such as boundary and the sense of place? • Did maps and mapping resolve or cause boundary and land use disputes? • Are there any changes in the property institutions that regulate resource access and claims? • Did maps and mapping change intra-community relationships?
SIT and NGOs	• How does an NGO decide to invest in developing an SIT component to their work? How does it decide on the choice of technology? • How do they sustain operating costs beyond initial investments? • Does the adoption or rejection of the technology affect relationships with donors? • Does adoption of SIT change the intra-organizational dynamics of an NGO? • Does a focus on participatory mapping affect the expectations of community members vis-à-vis NGO partners?

Grassroots Realities: SIT in Local Contexts

Mapping Power

Maps have long served to facilitate accumulation strategies that consolidate state control and works against the rights of local people (Harley 1992; Thongchai 1994; Brealey 1995; Escobar 1997). By the same logic, participatory mapping is viewed as a tool of empowerment and mediation for local communities, to re-insert their territorial claims onto 'empty' state maps. Mapping becomes a critical tool for negotiation with other groups, including neighbouring communities and the state. These activities occurred in the context of increased local activism, coupled with the opening of political space that followed the introduction of a new decentralization policy in Indonesia and the recognition of indigenous rights in the Philippines.

In Sarawak, the 2000 legal victory of Rumah Nor, an Iban village's claim against a tree plantation corporation (Majid Cooke 2003) energized communities across the state to organize and mapped their respective native customary lands. In a move to curb this rights-through-mapping legal power, in 2001 the Sarawak legislature passed the Land Surveyors law, which was designed to regulate community mapping. Nonetheless, as of 2005 more than 40 native customary land claims cases, based on community maps, have been filed in court – half of which were filed after the enactment of the law (Bujang 2005). Yet it is not clear if community maps will be admissible in future court proceedings to defend land rights of the indigenous Dayaks. Participatory mapping thus must be accompanied by legislative efforts to de-criminalize the activity; otherwise the extent of empowerment that the maps confer will be very limited.

Spatial information is useful for a variety of purposes at the grassroots level. Communities can better plan the management of their resources, monitor the implementation of development projects, and resolve resource conflicts within their own communities. Maps can give community members more knowledge about their resources, so they can respond better to problems. This potential is most visible in many communities that adopted SIT in developed economies such as the United States. For example, GIS has been an important tool for the Agricultural Land Preservation Board of Adam County in Pennsylvania to help residents recognize the rapidity of land-use change and the extent of threats to their resources (Dayhoff 2003). In Trinity County, California, Everett and Towle (2005) found that GIS helped local people to be more aware of their resources, which has led to greater sophistication in public discussions among communities and with public and private resource management. In these cases, mapping and working with maps enhanced community capacity in negotiating access to local resources, and increased community involvement in policy processes.

Others have cautioned, however, that mapping also helps outsiders gain knowledge for furthering their own interests. While mapping has enhanced tenure security (in Indonesia, Thailand, Cambodia, and the Philippines), it also benefited local governments by providing them with free information. Fox (2002) argues that if local people do not have control of their maps, they may not be any better off than they were before their lands were mapped. SIT data could contain information on valuable common resources such as birds' nests and honey trees that, if known by outsiders, could result in increased resource competition.

We also observed competing local/village institutions that oversee access to the maps and spatial information ranging from formal village governments, to traditional or customary institutions, to functional village committees. NGOs that initiate or sponsor community-mapping projects play key roles in influencing which actors exercise authority with respect to spatial information, and thus benefit from the adoption of SIT. For example, two NGOs in Indonesia chose divergent strategies. PPSDAK, a Kalimantan-based NGO chose to revitalize traditional customary institutions (*adat*), entrusting them with control of the maps (Lorensius 2003), while Koppesda, a Sumba-based NGO chose to support a functional committee on forest conservation and to bypass traditional leaders (Hardiono et al. 2005). The implications of these decisions can be far-reaching in the restructuring of power relations and property institutions that govern resource access and utilization.

Even if the community can control the maps, it is important to understand the multiple interests and actors found within communities; and the political and economic relationships between communities and other social actors. Within the client communities, mapping affects these relationships, causing new social stratification to emerge. In Malaysia, Mark Bujang (2005) noted a case in which entrusted community leaders colluded with a corporation, using community maps to support the corporation's plan to lease customary lands for an oil palm plantation.

Finally, we also need to be clear that not all maps and mapping activities are alike, with a spectrum of technological complexity that ranges from sketch maps to GIS. While paper maps are generally available to all at the local level, digital data presents a structural barrier that may prevent a large proportion of community members, as well as some NGO staff, from accessing the spatial data. In this case, determining who 'owns' the maps and the information they contain can be difficult. Reflecting on the case studies from Indonesia and from Cambodia, Hardiono et al. (2005) and Sarem, Ironside and van Rooijen (2005) noted that because the mapping facilitators and consultants that make community maps control the digital SIT databases, they effectively control access to the information they contain.

Impacts on communities' values

For many indigenous groups in Asia, the use of SIT in participatory mapping is primarily intended to 're-insert' their existence onto maps – to claim rights that had not been acknowledged by the state. Vandergeest and Peluso (1995) describe the process by which rights to resources are acknowledged by the state as territorialization. When resource rights have not previously been recognized and space has not yet been territorialized, mapping activities have greater impact on traditional ways of governing human environment interactions and seeing the world, than they do in communities where legal rights and territorialized space already exist.

We recognize, however, that changes in the sense of place and boundary conceptions are not exclusively caused by mapping activities, as they are also subject to changes in the political economic context, such as expansion of roads, markets, and state policies. For example, Setyowati (2006) documented shifting conceptions of rights and territoriality among the indigenous people of Siberut island, Indonesia through the eras of the timber industry in the 1970s; the national park/conservation movement in the 1980s and the early 1990s; and the decentralization since the fall of Suharto in 1998 (Setyowati 2006). Nonetheless, mapping accelerates these changes by facilitating direct influence of property institutions aligned with SIT. For example, if villagers engage in mapping in order to increase the security of their land claims, they need to follow through with land titling once they have mapped the land. But the land titling process is controlled by outside authorities, and has significant implications for their relations to the land, their neighbours, and their community. Mapping efforts initiated to recognize collective rights to land resources can lead to land privatization that is in the long run exclusive rather than inclusive.

We also recognize that mapping disadvantages nomadic groups that do not claim exclusive territories and therefore are generally not represented in the mapping process. In Malaysia, Indonesia, and Thailand, customary boundaries were traditionally flexible. These boundaries respond to changing needs within the community and extend across and overlap administrative boundaries, as well as the boundaries of neighbouring communities that may include nomadic groups. In communities who have mapped their territories, these boundaries have become less flexible and often cause disputes when they overlap with neighbouring boundaries.

Mapping can force communities to confront latent issues with regard to the management of natural resources. This can lead to new opportunities for consensus building, but it can also lead to conflict by making it harder to compromise positions, creating new disagreements within and between communities. One of the ironic effects of SIT observed in Cambodia is that mapping efforts initiated to resolve conflicts between local communities and government agencies resulted in increased conflict between and within villages (Prom and Ironside 2005). As long as boundaries remain fluid and flexible, defined only in a person's mental image of the landscape, conflicts between competing interests (within villages or between villages) can be minimized. Once boundaries are mapped, however, conflicting images of reality cannot be overlooked any longer and must be addressed.

Many participatory mapping proponents argue that they have no choice but to map. Today, the reach of mapping has extended to virtually every remote corner, leaving villagers with no 'exit option,' as they are already 'caught up in a mapping world'. They can refuse to map, but they cannot escape the implications of living in a world in which others will eventually map their lands. Mapping has become a precondition for protecting their territory and resources, as it is not possible to claim an unmapped area in contemporary politics. Even if you refuse to map within the boundaries of a protected territory, such as in a Native American reservation, the outer boundaries must be established and recognized. At the same time, villagers recognize that being included in official government maps can be as disadvantageous as being excluded from them (Majid Cooke 2001).

Furthermore, as SIT becomes a practical imperative, it ironically may disadvantage many small communities who do not have access to it. Likewise, resolving the conflicts caused by mapping draws attention to the importance of 'boundary' and 'territory' over other non-spatial aspects. This shift eventually makes SIT indispensable for asserting (and defending) communities' rights. In both Indonesia and Malaysia, many communities have realized 'the power of maps' and are anxious to have their resources mapped (Bujang 2005; Lorensius 2003). Yet the NGOs who assist in participatory mapping are unable to respond to all community requests for mapping. Communities that do not have maps become disadvantaged as 'rights' and 'power' are increasingly framed in spatial terms.

SIT and NGOs

We define non-government organizations (NGO) as organizations that work on a voluntary basis, rely on external funding, work with the poor and marginal members of society, have a small staff, and have a flexible, not-for-profit, independent and non-partisan nature (cf. Korten 1990). The urban and middle class nature of most NGOs as well as their dependence on funding from outside sources places their independence and performance in doubt.

Reflecting on their respective organizations' experience, participants of the workshop noted that decisions to incorporate SIT as an important component of NGOs' activities varied, but reasons external to the NGOs were at least as important as those from within. Donors, and how NGOs perceive donors' priorities, have a relatively large influence on many NGOs. Pramono (2005) describes how consultants from other international organizations – e.g. the East-West Center, the World Wildlife Fund, ICRAF, or the USAID-supported Biodiversity Support Program – proved to be instrumental for NGOS in Indonesia in their choice of mapping strategies. Similarly, Hardiono et al. (2005) describe how the shift from sketch mapping to GIS in Sumba, Indonesia was influenced by discussions with international actors. Donor's priorities, however, continue to evolve, and an NGO that received donor support to acquire SIT may not receive support to maintain the technology. It can also be difficult for an NGO to meet the timetables imposed by donors.

Success in using maps as tools for negotiating land rights in Indonesia and Malaysia has led to increased demand for mapping by neighbouring communities. This has created a shortage of technically trained people, and participants agreed that it is difficult to acquire and keep trained staff. There is also a gap in expectations and work culture between staff members trained in SIT sciences and those trained in social sciences that could lead to the separation of participatory mapping activities from the broader objective of NGOs (Hardiono et al. 2005).

Recognizing the potential socio-ethical impacts of SIT, there was a strong consensus among workshop participants that advocates of participatory mapping need a clear protocol to follow when introducing SIT into a village. This protocol should require outside actors to communicate clearly with each community prior to the mapping project. The NGO must clarify the purpose/objectives of collecting information, agree with villagers on what information can be mapped, and explain potential consequences of recording the community's spatial information on maps that can then be copied and distributed outside the community. Most importantly, outside facilitators must communicate that villagers can agree to accept or reject the mapping exercise.

Carrying out the protocol, however, is not sufficient in assuring that villagers would be aware of the full implications of mapping. In spite of the facilitators' efforts to organize meetings to discuss mapping issues, many villagers fail to attend the meetings (Bujang 2005). In some cases, the meeting schedules conflicted with the need of villagers to attend to their farms. In others, some villagers disagreed with the goals of participatory mapping and thus refused to participate in the conversation. Hardiono et al. (2005) and Sarem et al. (2005) highlight the problem of conceptual gaps between mapping facilitators or NGOs and villagers. In spite of the effort to consult with villagers and village leadership throughout the mapping process, the fact that many villagers had never seen or worked with maps made it difficult for them to fully comprehend the potential problems.

Unlike in North America, the use of SIT at the community level in Asia has largely been limited to producing one-time maps and neglecting the reality that working with spatial information is a process requiring revisions and changes. Thus far, little attention has been given to building local capacity to revise and re-map as circumstances change. Meeting this challenge will require not only building technical skills, but also skills for looking critically at context and for identifying factors needing response. Finally it will require sufficiently broad and keen ethical sensibilities to think through how changing practices set different directions for the community, carefully weighing options and their effects.

Summary

Our goal is to understand the social and ethical implications of the use of spatial information technology in community-based management, so that those who choose to use it to meet social objectives can do so wisely and with an understanding of the unintended consequences that may accompany its use. We seek to enhance the knowledge of the scientific community regarding the ethical, organizational, and power implications of spatial information technology, as well as to provide social activists with criteria for deciding whether they want to use this technology in their fieldwork.

The case studies we reviewed in our project confirmed that mapping, and working with maps, enhance community capacity to negotiate access to local resources. It develops technical and analytical skills in understanding both the immediate locale as a familiar place and its complex relationships to surrounding locales and regions. This wider perspective affords greater insight into current and likely patterns of interdependence, enabling better responses by communities to their own problems. As such, SIT is a useful capacity building resource for supporting the broader goals of community-based management.

It is important to understand that SIT comes in a variety of forms, and its conceptual and technical accessibility to participating communities could be uneven. Sketch mapping and 3D maps are easier to understand and are effective in engaging even illiterate villagers in conversations regarding natural resource management. But these maps are often considered to have limited credibility – a perception that markedly reduces their effectiveness when negotiating territorial rights with outside interests. However, efforts to 'formalize' SIT – away from sketch mapping toward technical cartographic mapping and GIS – could backfire. Indeed, in remote villages in Asia, adoption of technologically complex SIT could marginalize many of the targeted communities. Participatory mapping proponents therefore must strike a balance between being able to produce maps and spatial information that meet the cartographic convention, but that remain relevant to villagers in solving their immediate problems.

Reflections by practitioners as represented in the Chiang Mai workshop and the case studies, however, also identified several ironic effects of mapping that could undermine the goals of community-based management. While mapping is useful for bounding and staking claims to ancestral or traditional territories, it also facilitates a shift toward exclusive property rights and provides outsiders a legal means to gain access to common property resources. Common property resources are managed through rules and practices that include the control of knowledge about the location of valuable resources. By making knowledge accessible to all, mapping weakens existing common property management systems. Mapping generally promotes practices that shift attention and concern away from qualities of human/environment relationship to quantifiable limits on that relationship implied by boundaries/borders. The newly acquired authority to define and exert control over the use of space thus has begun to compromise the customary uses and governance it is intended to protect.

The impacts of SIT must also be seen in the context of how the participating communities are positioned in adopting the technology. Communities in the United States utilize SIT as a tool for capacity building. It is not intended to reform the structure of rights and access, but to facilitate communities in claiming those rights. By contrast, for many indigenous groups in Asia, the use of SIT in participatory mapping is primarily intended to claim rights that had not been acknowledged by the state. These new spatial practices, however, also bring about new ways of conceiving space and new patterns of relationship centred on spatially determined resources. The adoption of SIT and participatory mapping thus serves to infuse new values into user and user-affected communities. In indigenous groups and in smaller rural communities these new values can dramatically affect an array of existing paradigms, acting as catalysts for change in social organizations and in local dynamics of power and prestige.

The adoption of SIT and participatory mapping in Asia has increased the capacity of indigenous groups and local communities to assert territorial rights and to promote decentralization of resource governance and management. But the adoption of this technology has also increased the need for the further adoption of SIT by other rural communities, practically eliminating exit options. As participatory mapping practitioners in the workshop concluded, the more we map, the more likely it is that we will have no choice but to map. Yet we submit that this need not be seen as a caution against mapping, but rather as an injunction to develop critical clarity with respect to mapping based on a comprehensive understanding of both intended and likely unintended consequences of our actions. Resource managers who engage in mapping must do so with clear protocols for explaining these consequences to rural communities prior to the mapping exercise. Meeting this challenge will require not only building technical skills, but also transferring skills for looking critically at context and for identifying factors needing response. They must also work to establish a sustainable trajectory of community capacity building – a trajectory that insures continued, sufficient resources for the community to participate in negotiating political and economic relations that are continuously being transformed, sometimes in response to the adoption of SIT itself.

References

Brealey, K.G. (1995), 'Mapping them "Out": Euro-Canadian Cartography and the Appropriation of the Nuxalk and Ts'ilhqot in First Nations' Territories, 1793–1916', *Canadian Geographer* 39: 2, 140–156.

Brody, H. (1981), *Maps and Dreams: Indians and the British Columbia Frontier* (Vancouver: Douglas & McIntyre).

Brosius, P.J., Tsing, A. and Zerner, C. (1998), 'Representing Communities: History and Politics of Community-Based Resource Management', *Society and Natural Resources* 11, 157–168.

Bujang, M. (2005), 'Community-based Mapping', in Fox, J., Suryanata, K. and Hershock, P. (eds), *Mapping Communities: Ethics, Values, Practice* (Honolulu, HI: The East-West Center), 87–96.

Dayhoff, E.T. (2003), 'Adams County Pennsylvania Agricultural Land Preservation', Presentation to the Hawaii Agricultural Working Group (Honolulu).

Escobar, M. (1997), 'Exploration, Cartography and the Modernization of State Power', *International Social Science Journal* 49, 55–75.

Everett, Y. and Towle, P. (2005), 'Development of Rural Community Capacity Through Spatial Information Technology', in Fox, J., Suryanata K. and Hershock, P. (eds), *Mapping Communities: Ethics, Values, Practice* (Honolulu, HI: The East-West Center), 73–86.

Fox, J. (2002), 'Siam Mapped and Mapping in Cambodia: Boundaries, Sovereignty, and Indigenous Conceptions of Space', *Society and Natural Resources* 15: 1, 65–78.

Fox, J., Suryanata, K. and Hershock, P. (eds) (2005), *Mapping Communities: Ethics, Values, Practice* (Honolulu, HI: The East-West Center).

Hardiono, M., Radandima, H., Suryanata, K. and Fox, J. (2005), 'Building Local Capacity in Using SIT for Natural Resource Management in East Sumba, Indonesia', in Fox, J., Suryanata, K. and Hershock, P. (eds), *Mapping Communities: Ethics, Values, Practice* (Honolulu, HI: The East-West Center), 107–116.

Harley, J.B. (1992), 'Rereading the Maps of the Columbian Encounter', *Annals of the Association of American Geographers* 82: 3, 522–536.

Harris, T. and Weiner, D. (1998), 'Empowerment, Marginalization and "Community-integrated" GIS', *Cartography and GIS* 25: 2, 67–76.

Hershock, P.D. (1999), *Reinventing the Wheel: A Buddhist Response to the Information Age* (Albany, NY: State University of New York Press).

Illich, I. (1973), *Tools for Conviviality* (New York: Harper & Row).

Illich, I. (1981), *Shadow Work* (Boston: M. Boyars).

Korten, D.C. (1990), *Getting to the 21st Century: Voluntary Action and the Global Agenda* (West Hartford: Kumarian Press).

Lemke, J.L. (2000), 'Material Sign Processes and Emergent Ecosocial Organization', in Andersen, P.B., Emmeche, C., Finnemann, N.O. and Christiansen, P.V. (eds), *Downward Causation: Minds, Bodies and Matter* (Aarhus; Oakville, Conn: Aarhus University Press).

Lorensius (2003), 'Pemberdayaan Masyarakat Adat Melalui Pendekatan Pemetaan Partisipatif' [Enabling Customary Communities with Participatory Mapping Approach], paper presented at the workshop on Participatory Mapping: Opportunities and Challenges towards Spatial Sovereignty (Cipayung, Indonesia).

Majid Cooke, F. (2003), 'Maps and Counter-Maps: Globalised Imaginings and Local Realities of Sarawak's Plantation Agriculture', *Journal of Southeast Asian Studies* 34: 2, 265–284.

Mark, D.M., Chrisman, N., Frank, A.U., McHaffie, P.H. and Pickles, J. (n.d.), 'The GIS History Project', <www.geog.buffalo.edu/ncgia/gishist/bar_harbor.html>, accessed 6 November, 2000.

Peluso, N. (1995), 'Whose Woods are These? Counter Mapping Forest Territories in Kalimantan Indonesia', *Antipode* 27: 4, 383–406.

Poole, P. (1995), *Indigenous Peoples, Mapping and Biodiversity Conservation: An Analysis of Current Activities and Opportunities for Applying Geomatics Technologies* (Washington, DC: Biodiversity Support Program).

Pramono, A.H. (2005), 'Institutional Implications of Counter-mapping to Indonesian NGO's', in Fox, J., Suryanata, K. and Hershock, P. (eds), *Mapping Communities: Ethics, Values, Practice* (Honolulu, HI: The East-West Center), 97–106.

Prom, M. and J. Ironside (2005), 'Effective Maps for Planning Sustainable Land Use and Livelihoods', in Fox, J., Suryanata, K. and Hershock, P. (eds), *Mapping Communities: Ethics, Values, Practice* (Honolulu, HI: The East-West Center), 29–42.

Ribot, J.C. (2002), *Democratic Decentralization of Natural Resources: Institutionalizing Popular Participation* (Washington DC: World Resources Institute).

Rundstrom, R.A. (1995), 'GIS, Indigenous Peoples, and Epistemological Diversity', *Cartography and GIS* 22: 1, 45–57.

Sarem, K., Ironside, J. and van Rooijen, G. (2005), 'Understanding and Using Community Maps Among Indigenous Communities in Ratanakiri Province, Cambodia', in Fox, J., Suryanata, K. and Hershock, P. (eds), *Mapping Communities: Ethics, Values, Practice* (Honolulu, HI: The East-West Center), 43–56.

Setyowati, A.B. (2006), *Contested Terrains: Renegotiating Forest Access in Siberut Island, West Sumatra, Indonesia*, MA Thesis (University of Hawaii at Manoa, Department of Geography).

Shrader-Frechette, K.S. and Westra, L. (1997), *Technology and Values* (Lantham, MD: Rowman & Littlefield Publishers).

Sieber, R.E. (2000), 'GIS Implementation in the Grassroots', *URISA Journal* 12: 1, 15–28.

Sirait, M., Prasodjo, S., Podger, N., Flavelle, A. and Fox, J. (1994), 'Mapping Customary Land in East Kalimantan, Indonesia: A Tool for Forest Management', *Ambio* 23: 7, 411–417.

Sterritt, N.J., Galois, R., Overstall, R., Marsden, S. and Grant, P.R. (1999), *Tribal Boundaries in the Nass Watershed* (Vancouver, BC: UBC Press).

Tenner, E. (1996), *Why Things Bite Back: Technology and The Revenge of Unintended Consequences* (New York: Knopf).

Thongchai, W. (1994), *Siam Mapped: A History of the Geo-Body of a Nation* (Honolulu: University of Hawaii Press).

Urit, M. (2001), 'Land Surveyors Bill Goes Against NCR Lands', *Rengah Sarawak News*.

Vandergeest, P. and Peluso, N.L. (1995), 'Territorialization and State Power in Thailand', *Theory and Society* 24, 385–426.

Chapter 13

Competing and Conflicting Social Constructions of 'Land' in South Africa: The Case of and Implications for Land Reform

Brent McCusker

Introduction

This chapter examines how competing historically and socially constructed knowledges have affected the discourse on post-apartheid South Africa's land reform program. The knowledges include those of 'dispossessed communities', 'commercial' and/or 'white' farming groups. A wide range of interests and knowledges is contained within each of these groups. I make no attempt in this chapter to portray them as unproblematic constructions. I consider them 'knowledge-allies', i.e. groups who share a common knowledge or viewpoint. A second point of this chapter is to recount my experience of applying geo-spatial technologies (GSTs) in assessing land reform project performance. The contentious nature of the land reform debate led me to employ a seemingly more neutral arbiter to specific resource questions. I apply the findings of this exercise to the context of land reform and provide a few tentative recommendations for future research and policy.

This analysis draws on a political ecology approach to understanding human-environment interactions. In a political ecology approach, social, political, environmental and economic contexts temper land use decisions made across all scales and through time (Campbell and Olson 1991). I draw upon two recent case studies northern South Africa to demonstrate how competing groups construct their own knowledges to support their main claims and how I used GIS and remote sensing to investigate some of these contradictory claims.

In the second section of this chapter, I briefly outline the theoretical and conceptual frameworks that bind the central argument to the case studies and the conclusion. This section represents my particular perspective on the issue. I briefly examine land reform policy in the third section and discuss the study sites in the fourth section. Two competing knowledges are reviewed in the fifth section. I draw together the major arguments and comment on the use of knowledge to advance particular land agendas in the conclusion.

Theoretical Framework

While exceptionally diverse, studies that have drawn on political ecology as an explanatory framework are bound by the notion that human interaction with the environment is constructed politically and socially, across scale and time, and often riddled with conflict. Robbins reviewed and synthesized a range of 'political ecology' studies and describes the field as consisting of:

> empirical, research-based explorations to explain linkages in the condition and change of social/environmental systems, with explicit consideration of relations of power ... it is a field that stresses not only that ecological systems are political, but also that our very ideas about them are further delimited and directed through political and economic processes (2004: 12).

Ben Page described five traditions within political ecological studies in a 2003 article entitled 'The Political Ecology of Prunius Africanus' that I find useful for positioning this chapter. He noted that political ecology's first two traditions include writings that engage ecological concepts for the understanding of politics and the 'political wing of the environmental movement' (Page 2003: 358). Given that both of these conceptualizations of political ecology are vastly different from those represented by this book, I focus on the other three. The third tradition is rooted in a Marxian 'political-economic analysis of the relationship between society and nature' (358), while the fourth tradition 'emerged from economic anthropology ... and was concerned with the relationships between the physical environment, production, resource ownership, and the distribution of people' (358). A cultural ecological approach, the fifth, is 'explicitly concerned with questions of scale and biophysical processes and not purely political economy' (359). This chapter draws on the third and fifth traditions heavily and includes the fourth as it illustrates how shifts in production are often linked to shifts in environmental 'knowledge'. While both the Marxist and biophysical traditions within political ecology have tended to emphasize different aspects of the human-environment relationship, they both engage 'nature' as socially constructed and contentious.

My perspective originates from this and broadly related literatures. I find particularly relevant the notions that *all* knowledge is socially constructed (Latour 1987, 1993). Further, 'nature' is used as a container for the 'natural' environment and human relations within it, which are socially constructed to foment capitalist penetration and utilization of environmental resources (see Castree 1995, 2003). When I apply these claims to my own work, I find that the technicist turn in geography and its recent integration with political ecology has led to claims of the emancipatory potential of geospatial technologies (Weiner, et al. 1995) that have not often 'levelled' the field by interrogating *both* technical/expert knowledge *and* local knowledge as socially, culturally, and politically constructed (for a good review of this claim, see Robbins 2003). Such constitutions of knowledge are historically contingent and penetrate *within and between* rather than *down to* 'communities' in South Africa (James 2000). In this respect, communities and individuals within those communities are dynamic actors in the constitution of these knowledges, not simply passive observers. This perspective facilitates an interrogation of the creation of categories (natives, progressive farmers, sustainability, soil erosion, land degradation) through which land use control is eventually affected. This is what Robbins referred to as the 'simultaneous examination of the partialities of *all* knowledge' (Robbins 2003: 235).

The construction of these knowledge-allies are fleeting, contested, and fraught with contradiction. Local communities are not collections of common political/cultural/social interests, especially given the apartheid government's abuse and misuse of such constructs. Geo-racist labour and space allocation was a major driving force of colonial and apartheid rural spatial engineering (Platsky and Walker 1983). The idea that these 'communities' would now produce a 'common' interest formed from a common 'local knowledge' simplifies, if not ignores, both pre- and post-apartheid land struggles, and undermines independent political action, subjugating the individual to the yoke of the community.

These community 'knowledges', on the other hand, can also be inclusive and give shelter to a variety of individuals. Individuals often pursue collective action in the land reform process as a type of insurance against the expense and risk of such activities. Were individuals to 'go-it-alone', it is likely that very few land reform claims would have been settled. Collective action through collective stories can be a powerful force of change within the land reform program since oral histories are a cornerstone of the claim process (pers. Comm. DLA official, June 2006).

Having found these competing constructions rich in contradictions, I turned to geo-spatial technologies (GSTs) to examine the epistemological basis for such knowledge claims. A growing number of political ecologists (and closely aligned researchers) are engaging technology to gain an additional perspective on nature-society interactions in the landscape (e.g. Fox et al. 2003 and this edition; Robbins 2003; McCusker and Weiner 2003; King 2002; Walker and Peters 2001). These political ecologists are not alone in their interest; issues regarding the integration of geo-spatial technology (GST) and social science have been investigated by of a wide variety of researchers in literature often referred to as public participation GIS (Pickles, 1991, 1995; Sheppard, et al. 1999). These researchers, however, infuse a healthy dose of skepticism (Turner and Hiernaux 2001; Turner 1999), as geo-spatial tools have been used in the past to enforce domination. Much of the recent integrationist literature embraces the notion that the emancipatory potential of GSTs does not preclude their use for empire building and clandestine state surveillance of citizens.

From this theoretical and methodological perspective, I turn to the second part of the title of this chapter, the 'land', or more accurately land reform policy that has been the object of competing social constructions. The struggle over the definition, implementation and control of the discourse over land has been contentious, to say the least, over the past fourteen years in South Africa.

Land Reform Policy

South Africa's official land reform program began well before the ascendancy of the ANC to office in 1994. During the lead up to democratic elections, the deKlerk regime slowly began lifting the most heinous apartheid edicts. A key point in redressing apartheid injustices came in 1991 with the passage of the *Abolition of Racially Based Land Measures Act* (Act 108 of 1991). This act repealed the most notorious laws including the *Natives Land Act* (1913), the *Native Trust and Land Act* (1936), the *Group Areas Act* (1950) among others. During the negotiated settlement period, additional redress for the years of discriminatory legislation came in 1993 with the passage of the *Provision of Certain Land for Settlement Act* (Act 126 of 1993). This act established certain conditions for supporting transfer of land, although the process was more firmly established in 1994 with the passage of the *Restitution of Land Rights Act* (Act 22 of 1994). This process was wrought with contradictions, but no so more obvious than the gradual acceptance of the ANC market-based principles for land reform.

In the land reform program from 1994–2000, policy advisers welded together the market-based mechanism to a 'pro-poor' approach to land redistribution in their recommendations to government. International experience in land reform was employed in an effort to preclude policy makers in South Africa from the same mistakes repeated in Africa, Latin America, and East Asia. Bernstein (1999) notes that so-called 'reformed' apartheid-era academics felt certain that the market-driven policies that worked so well for white South Africa could be transformed into 'dispossessed-friendly' policy. South Africa emerged from apartheid at the heyday of World Bank and IMF restructuring programs that emphasized 'getting the markets right'. International agricultural economists stressed the need for market driven reforms. This perspective is represented in the work *Agricultural Land Reform in South Africa*, which stresses the need for market-based land reform. While communal and group projects are discussed as is the need to satiate the appetite of the poor and landless for land, the collection clearly outlines the evidence in favour of market-based, business-oriented reform.

The Department of Land Affairs allowed barely five years for the implementation of the market-based/ pro-poor approach to land reform before condemning it as ineffective (Department of Land Affairs 1999). Rather than re-examining the 'market-based' mechanism, it was the 'pro-poor approach' that was to be scrapped. In 2000, the Land Redistribution for Agricultural Development framework was launched, emphasizing the importance of supporting emerging 'entrepreneurial' black farmers. While not completing abandoning the poor, entry requirements made their participation much more difficult. This policy shift was not the result of a long process of monitoring and evaluation and adjustment of the pro-poor approach; counter-narratives to land reform emerged quickly that prompted South Africa land planners to re-examine their own agenda.

After a brief description of the study sites, I document two narratives about land reform in attempt to understand why government's land reform has been so slow to achieve results. I interrogate the construction of knowledge in this section to address the last portion of the title of this chapter – the 'implications for land reform'.

Land Reform Study Sites

The farms on which I conducted research were created under the Communal Property Associations Act (No. 28 of 1996). This act facilitated land redistribution through the pro-poor framework and created a legal institution, the Communal Property Association (CPA), to promote land ownership by historically disadvantaged groups, foster gender equity, and provide opportunities for landless rural people to practice agriculture and/or animal husbandry. Each association had a constitution, a business plan, and a democratically elected governing body. I assessed livelihood and land use changes on five of these associations over the course of a year in 1999 and again in 2001. I conducted follow-up on all five and reviewed a sixth in May–June 2006.

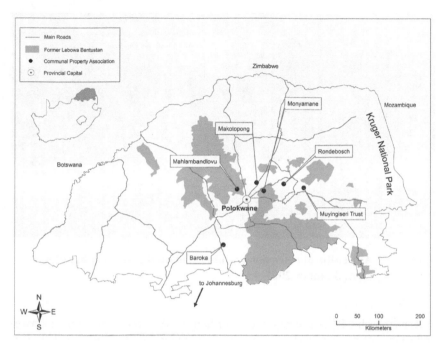

Map 13.1   Location of the study sites in Limpopo

The Mahlambandlovu CPA is located approximately 20km from Polokwane just across the border of the former Lebowa homeland in the highveld. The CPA purchased several farms from a (white) large-scale commercial farmer 1997 and formally established itself as an operational CPA in 1998. During the initial study period, the CPA had a small truck, two tractors, and an array of farming implements with which they raised chickens and cattle. The farmland for the CPA is located in a lower-potential zone for farming, receiving lower rainfall (400–600mm) and with generally less productive soils. CPA membership stood at 396, but the secretary estimated that there are fewer than 160 active members. After three months of participatory fieldwork, 120 members were interviewed.

**Figure 13.1   Agricultural extension training plot on the Mahlambandlovu
CPA, January 2001**

The Rondebosch CPA is situated in the northern Drakensburg Escarpment between
Polokwane and Modjadjiskloof. The thirty members of the CPA resided in the
Sekgopo community, which lies just to the northeast of the actual CPA land. The
farm land of the CPA was in a higher-potential area that receives more rainfall
(600–800mm) and has more productive soils, especially when compared to the
Mahlambandlovu CPA.

The Monyamane CPA is located 15km north of the University of Limpopo. The
CPA was established in 1997 and officially serves 201 households. The farms that
constitute the CPA lie in exceedingly steep terrain and as such, accessibility is severely
limited with much of the land was accessible only by foot. Many members of the CPA
reside just to the northwest of the CPA itself.

The Baroka CPA lies between Makopane and Roedtan on the N11 highway. It was
the second smallest of the CPAs visited in both size (5ha) and membership (41 people).
The farm was established in 1997 and finalized in 1998 as part of a larger white farm
that was sub-divided and sold. The members constructed a small irrigation scheme that
provided water to a small field producing vegetables. The other portion of their land
has been set aside for bean, maize, and tomato production.

The Muyingiseri Trust was located northeast of Tzaneen on the Phalaborwa (R71)
road. The CPA was established in 1997 and finalized in 1998. The trust is the smallest in
membership of all of the farms visited (39 beneficiaries). At the time of interviews the

road into the farm was completely inaccessible, also due to the 2000 floods. Production on the farm was limited to a small stand of tree crops and a few vegetable gardens. The farm was auctioned in 2006 to pay for an overdue electricity bill.

In 2006, I investigated a sixth CPA, the Makotopong CPA while conducting research on rural township growth in central Limpopo. Many people in Makotopong township were forcibly removed from the Roodewal Mission north of Polokwane in June 1967 (local respondents; SPP 1979). A group of removed people were approved for land redistribution through the CPA program in 2002. During a survey of the community, the topic of the CPA arose. Having already investigated five CPAs, I pursued a line of questioning about the Makotopong CPA as well.

In the next section I outline two competing narratives about land reform that directly affects these six CPAs. Their stories are far from straightforward and often make the local and national press for both positive and negative reasons. The narratives explaining the experience of these projects differ sharply depending on whom one questions.

**Figure 13.2   Destruction of farmland at the Majakame Rondebosch
CPA, March 2000**

Contentious Narratives

The differing narratives expressed in South Africa regarding land reform are held emphatically by their proponents. These contentious narratives result largely from the very different perceptions of history. I attempt to compare each of these dominant knowledges through deconstruction in this section. Each knowledge contains certain accepted wisdoms that lead to the contentious nature of the land reform process in South Africa. I begin by examining the use of the terms 'dispossessed' and "community" to describe the beneficiaries of reform and summarize the main grievances that white farming communities holds against the land reform process.

Land reform for whom? 'Dis'-ing the dispossessed?

The notion that groups of dispossessed people would band together around common interests, usually farming, was and is still widespread in land reform policy. Group purchase and production was enshrined in both the Communal Property Association Act and by the Settlement/Land Acquisition Grant mechanism. Group claims are also accepted in restitution as well as group tenure arrangements.

Walker (2003) challenged such a focus. She wrote that 'for many land activists, indeed for many South Africans, [the] history of dispossession is constitutive of the social and political identity of black people as a group, inclusive of people who may not themselves have experienced land loss or forced removals (many even have benefited from such processes in the past)' (Walker 2003: 5). The knowledges and experiences of these communities, in many cases, are often ones of convenience and expediency rather than the sharing of any collective knowledge or feeling of community. Unfortunately, the construction of land reform around these ideas has led to many problems. Turner and Ibsen (2003), in a review of land reform, reported that 'there was little group cohesion or identity' on many pro-poor projects (Turner and Ibsen 2003: 15). I have previously highlighted the problem of a lack of group cohesion on communal farms in Limpopo (McCusker 2002, 2004).

One of the issues that I have observed over the past few years researching land reform projects is that method through which people were recruited for redistribution projects has led to a great deal of conflict. Because land redistribution projects in the 1994–2000 period were not aimed at groups of individuals who had a similar claim, but rather a wide cross-section of 'dispossessed people' who met certain income and race qualifiers, the 'community' that emerged in the redistribution schemes scarcely represented anything communal at all.[1] Thus, many beneficiaries in the CPAs investigated were little more than individuals in a 'rent-a-crowd' scheme whose signatures were utilized by a small interest group to obtain land (Turner and Ibsen 2000: 15). Their rights, and the risks of joining the schemes, were poorly – if at all – expressed to them. The main object of their signature was to obtain their R16,000 needed in order to purchase farm land. This sets up the first problem in defining 'dispossessed people'. Assuming that 'dispossessed' people would simply band together in the name of communal production without consideration of existing

---

1   This statement does not necessarily apply to land restitution projects that operate under very different dynamics from land redistribution projects.

political and social relationships was one of many mistakes built into South Africa's land reform program.

A second problem arose when discussing agricultural production. While most of the people participating were indeed dispossessed, the more significant question was 'from what were they dispossessed'? Discussions with the members of study groups indicated that they were forcibly removed under apartheid as part of the regime's obsession with racial segregation and rural planning. Such planning was never designed for anything more than extracting labour from the bantustans. While these beneficiaries may have had access to better farming land prior to their removal, many were neither skilled at the *business* of farming nor in any position to mobilize the labour necessary for farm production. These individuals had been dispossessed of their *familial labour* (that was required in the mines of Johannesburg) during apartheid not any significant farming income. This latter dispossession took place long before the members of these associations were born or had formed separate family units (Bundy 1979). Except for the oldest members of the associations, very few beneficiaries recounted having any knowledge of the business of farming. This finding does not indicate that the widespread dispossession and alienation did not occur under the apartheid, rather that it occurred so far in the past that the younger members of the associations never grew up in the peasant economy. Assuming 'dispossession' took the form of land alienation over labour alienation was a second mistaken assumption built into the first land redistribution framework.

The third issue facing land redistribution projects is that the concept of the 'dispossessed' is often used in such a way as to refer to an undifferentiated group of 'powerless' or 'poverty-stricken' individuals, families, and communities. The romantic notion of empowerment in modern South Africa has been replaced with the stark reality that no matter how well designed, a 'strong-man' (and/or a 'strong-woman') often emerges with aspirations beyond local economic development. The use of 'dispossessed' and 'community' were/are often hijacked to serve the interests of a narrow power elite.

Discussions with members of Makotopong CPA, just north of Polokwane, revealed that internal fighting over farm management ended all hopes at producing anything from the land that the 'community' was awarded. Indeed, at the 'betterment town' where most members of the CPA resided, there was a clear power struggle between competing individuals and groups who were jockeying for control. Being former South African Development Trust land, the area was not supposed to have a 'tribal' authority structure, yet a headman claimed on repeated occasions to be the 'chief' of Makotopong, even embellishing his office door with a hand written placard exclaiming 'CHIEF'. When the issue of a land claim pending on the former farmland was raised during a 'community meeting', the 'chief' and his secretaries shutdown all lines of inquiry. Informants separately advised us that this 'chief' was heavily involved in both the land redistribution project (the Makotopong CPA) and the land *restitution* claim. This fact was confirmed by cross-checking the list of CPA members and the restitution claimant list. Discussions were less detailed and relied only on qualitative methods, however, beneficiaries of the CPA reported that 'nothing is happening on that farm. All the pipes and tractors have been stolen and you will not find anyone to talk to about it. The chairman of that farm doesn't even live here' (pers. Comm., anonymous CPA beneficiary, June 2006). Further discussions repeated the same story about the collapse of the CPA due to gross mismanagement and in-fighting of the farm members. A visit to the farm confirmed that indeed there was no productive activity being undertaken.

Competing constructions of knowledge at the community level foment discord even where communities do share a common history of dispossession. On farm after farm, members of redistribution projects struggled to eke any production at all out of their farms. On Monyamane, Rondebosch, Makotopong, Baroka, and Muyingiseri land redistribution schemes, similar stories of conflict rose to be one of the more important factors leading to farm difficulties. Clearly, the large communal projects of South Africa's first period of land reform were overwhelmed by internal politics. The basic constructs upon which they relied to build their narrative, dispossession, community and production, are problematic at best. As detailed in the next section, the misuse of this set-back has threatened the entire redistributive land reform program.

Against land reform: commercial and white farming narratives

South Africa's white farming community was deeply frightened by the events of 2000 in Zimbabwe. Robert Mugabe's land reform initiative in that year violently forced white farmers off their holdings and replaced them with so-called 'war-veterans'. The reverberations south of the Limpopo were strong and swift.

Many in South Africa's white farming community felt antagonized even before Mugabe's land grab in Zimbabwe. While maintaining a willing-buyer, willing-seller approach, government has been quick to point out its willingness to use expropriation against "obstinate" farmers (for examples of news articles that remind readers of the government's option to use expropriation to procure individual farms in contentious land cases, see – *Mail and Guardian*, 02/20/2001; 03/20/2001; 10/20/2003; 01/08/2004; 02/07/2006). Additionally, press reports fuel a sense of 'siege' by pointing out the widespread discontent among landless groups and activists over the pace and progress of the current reform program. Headlines such as 'People running out of patience for land, says MP' and 'Claimants seek 20 per cent of SA farmland' engender fear about the future across many white farms (*Mail and Guardian*, 05/19/2006 and BBC News Website 01/07/2004).

As a result of this meta-narrative over land, and in an effort to protect their vested interests, white farming interests have been quick to point out each and every failure of land reform. Anecdotes of failures of farming on transferred lands are polemically represented in Phillip du Toit's 2004 book 'The Great South African Land Scandal'. Chapter after chapter recounts the 'arrogance and incompetence' of the beneficiaries of land transfers and the ignorance or ideological intransigence of government. Unsurprisingly, the book does not recount the millions of Rands in subsidies that accrued to each of these farms, nor the incalculable costs of the heavily protected markets that privileged the white owners under apartheid.

The antagonistic stance of many actors in the white farming community revolves around several key claims. The following examples have been culled from literature where such statements were attributed to one or more parts of the white farming community. They are not necessarily representative of the views of each author:

1. Blacks are not efficient or productive farmers (Ramutsindela 1998: 160).
2. The removal process under apartheid was legal (*Mail* and *Guardian*, 23 June 2006).
3. The transfer of land to the majority will lead to collapse of farming and/or the export sector (du Toit 2004).
4. Projects transferred thus far under reform are largely failures (du Toit 2004).
5. Corruption and graft are widespread in reform and thus, the beneficiaries do not actually benefit from land reform (pers. comm. Commercial farmer in Limpopo, January 2001).

These knowledges have been repeated by white farmers across Limpopo to me over the past seven years. A 'beautiful' and 'well-run' farm becomes defunct and derelict after restitution or redistribution due to 'incompetence' on the part of government and beneficiaries. By contrast, white farmers go to great lengths to show how their operation is efficient, productive and beneficial to society. In Limpopo, white farmers rely heavily on the economic efficiency argument. The dense concentration of bantustans in Limpopo combined with a high number of forced removals leaves many white farmers in a tenuous legal position regarding their ownership of the land, thus they have sought to minimize the 'rights' to land argument in favor of the 'efficiency' argument.

Figure 13.3   Ample evidence? The collapse of Muyingiseri Trust led the Tzaneen Municipality to sell the farm for a tenth of its original purchase price. This is the road to the farm in March 2000

Combined with the fear of losing their land, many white farmers are bitter over the loss of subsidies, arguing that farming provides the country with food and foreign currency. They argue that the cutting of subsidies was politically motivated by a hostile ANC. As a result, these groups tend to be hostile to the entire land claims process. Theo de Jager, head of the local commercial farmers association, was quoted regarding a pending claim in the central part of the province: 'we will oppose any expropriation in court, because the Makgobas [the local black group] never lived on most of the land under claim. And, in those parts where the Makgobas lived, our research tells us that they were removed in about 1895, which is many years before the 1913 cut-off date for claims' (*Mail* and *Guardian*, 06/23/2006). The battle of competing knowledges is especially intense in high-profiles cases such as this one.

The evidence most often fronted in their objection to all land reform is the inability of the beneficiaries to "effectively utilize" the land. Unfortunately, as shown above, the forces opposed to reform have ample ammunition to work with.

Land reform and GSTs: just another 'layer' of knowledge?

Given the layers of conflict and competing knowledges on land reform projects, I utilized GSTs in order to provide another socially constructed vantage point from which to observe the outcomes of South Africa's land reform program. The advantage of using GSTs is that they introduce photographic evidence (both air photos and satellite imagery) to understand and address specific claims. The disadvantages include the fact that the social and political processes behind observations of the landscape are not apparent from the imagery, giving the detractors of land reform incomplete, but popularly accepted, evidence that land reform is ineffective.

To analyze land use changes on the CPAs, I obtained six Landsat Thematic Mapper images covering the communal farms for the years 1988 and 2000/2001. I digitized select areas of the images to determine land cover, land use, and change. CPA beneficiaries discussed their ideas and perceptions of land use and cover change through a quantitative survey, group project meetings, informal discussions in the field and intensive individual interviews. Respondents were given detailed explanations of the GIS derived maps and asked to describe why changes happened in certain areas. The maps were sufficiently detailed for respondents to explain specific land uses and reasons for change. The large sample size of the quantitative survey allowed for verification and comparison of perceptions of change.

The study of land use and land cover change included and analysis of production intensification, dis-intensification and extensification, where intensification refers to the shift in land use from a state where value of product per land unit increases, while extensification refers to shifts in land use where the value of product per land unit decreases.

On the Rondebosch, Muyingiseri, and Monyamane CPAs widespread dis-intensification and abandonment occurred after the farms were transferred to beneficiaries. Regarding Mahlambandlovu CPA, land use change consisted largely of extensification, although some intensification was noted in follow-up research in 2001. In the western area of the CPA land that had been used in 1988 as farmland has since become grassland.

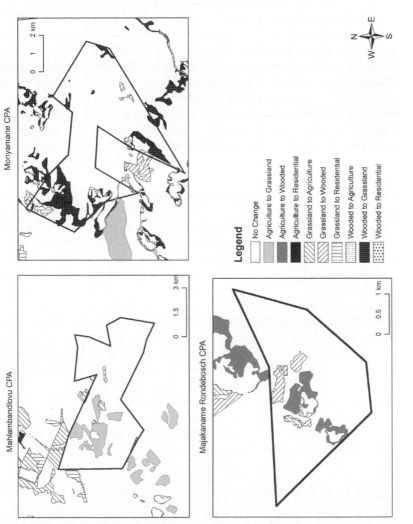

Figure 13.4　Land use changes on three Communal Property Associations, 1988–2000

It would be erroneous to end the analysis using GSTs at this point. As a socially constructed knowledge, GSTs are limited by the user and his/her assumptions about the landscape that they are viewing (see especially Walker and Peters: 2001 and Weiner, et al. 1995), as such, certain critical processes may be 'hidden' from the GST expert. I identified one such example near one of the larger CPAs. During a transect walk a beneficiary noted that I had classified a certain plot of land as grassland in the 1988 imagery and as agricultural land in the 2000 imagery. He suggested that there must be an error as the community had used that plot in 1988 as maize fields (agriculture). On returning from the transect walk and reviewing the imagery, I determined that there was no possible way that the plot was previously under agricultural use. During an informal interview of a few people near the spot, many individuals were insistent that the land 'had always been used for maize fields, including 1988'. Even after having shown and explained the imagery and reconfirmed that the plot in the image was the contested plot on the ground, they were convinced that the plot was used for maize fields in 1988. The seeming conundrum was addressed after the meeting. The next week while doing follow-up work in an area nearby, three people from the community meeting approached me. An older woman explained that the area *had been* grassland from the early 80s until about 1996 or 1997. Several members of the community were occupying the land in hopes that the local chief and/or transitional local council would grant them permanent rights to the land as a first step toward resolution of a formal land claim. Part of the basis of their claim was that they had been consistently working the land for 'many years'. The informants maintained that if it was shown that the land was not used as recently as 1988, they feared that their claim to the land would not be accepted, thus the dispute over land use.[2]

The use of the modifier 'hidden' here refers to a very specific point of view, that of the 'expert.' This knowledge was only hidden because informants did not want to make me privy to it; there is simply no way to 'see' this through any amount of image interpretation. In this example, it is clear that a political ecology analysis of land use without the use of GSTs may have never identified the power struggle over this particular piece of land, but it also emphasizes the need to treat all knowledge as socially constructed, including that derived through the use of GSTs. The important point here is not that one approach is more or less valid, but that combining 'knowledges' helps to evoke a more vivid portrayal of the political and ecological processes at work in contested landscapes.

Conclusion: The Land Reform Quandary

The point of this chapter has been to examine competing knowledges that have jockeyed to affect (or dis-affect) land reform in post-apartheid South Africa. I have reviewed two of the knowledges on land reform that I have encountered over seven years of research in Limpopo. My attempt to use geo-spatial technologies to address land reform problems on six transferred farms raised as many questions

---

2   The actual location of this plot and farm name is withheld to protect the privacy rights of the informants.

as it answered. I found that land has largely become abandoned or used for much less intensive farming. I highlighted two competing narratives often employed to influence policy and showed that neither narrative quite matched its claims.

Given the immense pressure to resolve the current impasse over land reform, academics and government planners may be tempted to short-cut the contentious debate described in this chapter through the employment of such technologies. These technologies, however, are not neutral arbiters. GSTs are socially constructed with very specific knowledge (data) 'quality' requirements. Additionally, this knowledge (data) has to meet rigorous formatting requirements, the most basic of which is that it is Cartesian. Vast stores of knowledge about land, land use, and land history defy such formatting and quality standards. Additionally, no amount of expertise or imagery can uncover the multitudes of social processes functioning in the landscape. Government planners and academics who plan to use GSTs to address contentious land issues will be well-served to keep this in mind as they design studies/policies.

To the broader question of land reform in South Africa, the contentious knowledges outlined in this chapter will continue to jockey for influence over the land reform program, however, the government ultimately bears the responsibility for the successes and failures of the program. It must navigate the intense pressure from both sides of the land reform debate – the dispossessed and the entrenched farming interests. To do so it must not accept either 'local' or 'expert' knowledge as unproblematic nor can it discount the knowledge of white/commercial farmers. In the past, the government has attempted to placate all interests and in the process has satisfied none. Clearly, the large communal production projects experienced serious problems and needed to be re-examined. Both the pro-poor and entrepreneurial approached have not met planners expectations and have been exceptionally costly. Conventional, academic, and government wisdom in South Africa suggests that task of the Department of Land Affairs is to plan land reform that satiates the appetites of the landless masses while not 'scaring-off' investors or plunging South Africa into a Zimbabwe style crisis. The knowledges examined in this chapter are all jockeying to reshape land reform in South Africa to increase their power over the process. The government must be willing to interrogate all claims made about land reform, not just the politically or economically popular ones. The question for future research will be to explore how government can best arbitrate between the competing socially constructed, and very entrenched, forms of knowledge.

Acknowledgements

I owe a deep debt of gratitude to the many people who helped me during this research. I would like to acknowledge Drs. David Campbell and Daniel Weiner for thier support and encouragement, Dr. Paul Fouche, Parvin Shaker and Marubini Ramudzuli for their invaluable assistance during my visits, and the NSF, NASA, the WVU Faculty Senate Office, and the Office of International Programs, WVU for their generous financial support. Finally, I reserve my most generous thanks for my family, Eileen, Grace and Ian, for their patience during all those long summers of fieldwork.

References

BBC News (2004), *Claimants Seek 20 per cent of SA Farmland*, accessed at <http:// news.bbc.co.uk> on 7 January at 14:41 GMT.

Bernstein, H. (1999), *Social Change in the South African Countryside? Land and Production, Poverty and Power*, Paper Presented at the Conference 'Land and Agrarian Reform in South Africa: Successes, Problems, and the Way Forward', 26–28 July, 1999, Pretoria.

Campbell, D. and Olson, J. (1991), *Framework for Environment and Development: The Kite*, CASID Occasional Paper No. 10 (East Lansing: MSU Press), 24.

Castree, N. (1995) 'The Nature of Produced Nature: Materiality and Knowledge Construction in Marxism', *Antipode* 27: 1, 12–`48.

Castree, N. (2003), 'Commodifying What Nature?' *Progress in Human Geography* 27: 3, 273–297.

Du Toit, P. (2004), *The Great South African Land Scandal* (Legacy Publications, Centurion).

Fox, J., Rindfuss, R.R., Walsh, S.J. and Mishra, V. (eds) (2003), *People and the Environment: Approaches for Linking Household and Community Surveys to GIS and Remote Sensing* (Boston: Kluwer Academic Publishers).

James, D. (2000), 'Hill of Thorns: Custom, Knowledge and the Reclaiming of a Lost Land in the New South Africa', *Development and Change* 31, 629–649.

King, B. (2002), 'Towards a Participatory GIS: Evaluating Case Studies of Participatory Rural Appraisal and GIS in the Developing World', *Cartography and Geographic Information Science* 29: 1, 43–52.

Land Affairs, Department of (1999), *Annual Report 1999* (Pretoria: Department of Land Affairs).

Latour, B. (1987), *Science in Action: How to Follow Scientists and Engineers through Society* (Cambridge: Harvard University Press).

Latour, B. (1993), *We Have Never Been Modern* (Cambridge: Harvard University Press).

Mail and Guardian (2001), 'Farmer Battles over Boomplaats Order', 20 March.

Mail and Guardian (2003), '"Vital" to Speed Up Land Reform', 20 October.

Mail and Guardian (2004), 'SA Land Restitution: Fears of Another Zimbabwe', 8 January.

Mail and Guardian (2006), 'Farmers Urge Rethink of Land Expropriation', 7 February.

Mail and Guardian (2006), 'People "Running Out of Patience" for Land Says MP', 19 May.

Mail and Guardian (2006), 'Move, or Be Moved', 23 June.

McCusker, B. (2002), 'The Impact of Membership in Communal Property Associations on Livelihoods in the Northern Province, South Africa', *GeoJournal* 56: 2, 113–122.

McCusker, B. and Weiner, D. (2003), 'GIS Representations of Nature, Political Ecology, and the Study of Land Use and Land Cover Change in South Africa', in Zimmerer, K. and Bassett, T. (eds), *Political Ecology: An Integrative Approach to Geography and Environment-Development Studies* (New York: Guilford).

McCusker, B. (2004), 'Land Use Change on Recently Redistributed Farms in the Northern Province, South Africa', *Human Ecology* 32: 1, 49–75.

Page, B. (2003), 'The Political Ecology of Prunus Africana in Cameroon', *Area* 35: 4, 357–370.

Platsky, L. and Walker, C. (1983), *The Surplus People: Forced Removals in South Africa* (Johannesburg: Ravan Press).

Ramutsindela, M. (1998), 'Compromises and Consequences: An Analysis of South Africa's Land Reform Program', *The Arab World Geographer* 1: 2, 155–169.

Robbins, P. (2003), 'Beyond Ground Truth: GIS and the Environmental Knowledge of Herders, Professional Foresters, and Other Traditional Communities', *Human Ecology* 31: 2, 233–253.

Robbins, P. (2004), *Political Ecology: A Critical Introduction* (Malden: Blackwell).

Sheppard, E., Couclelis, H., Graham, S., Harrington, J., Onsrud, H. (1999), 'Geographies of the Information Society', *Int. J. Geographical Information Science* 13: 8, 797–823.

Turner, M. (1999), 'Merging Local and Regional Analyses of Land-Use Change: The Case of Livestock in the Sahel', *Annals of the Association of American Geographers* 89: 2, 191–219.

Turner, M. and Hiernaux, P. (2002), 'The Use of Herders' Accounts to Map Livestock Activities Across Agropastoral Landscapes in Semi-Arid Africa', *Landscape Ecology* 17, 367–385.

Turner, S. and Ibsen, H. (2000), 'Land and Agrarian Reform in South Africa: Research Report No. 6' (Bellvile, SA: Programme for Land and Agrarian Studies).

Walker, C. (2003), *The Limits to Land Reform: Reviewing the Land Question*, Paper Presented at the African Studies/History Seminar, University of Natal, Durban, November 2003.

Walker, P. and Peters, P. (2001), 'Maps, Metaphors, and Meanings: Boundary Struggles and Village Forest Use on Private and State Land in Malawi', *Society and Natural Resources* 14, 411–424.

Weiner, D., Warner, T., Harris, T., and Levin, R. (1995), 'Apartheid Representations in a Digital Landscape: GIS, Remote Sensing and Local Knowledge in Kiepersol, South Africa', *Cartography and Geographic Information Systems* 22: 1, 30–44.

Index